Let us derive
$e=mc^2$

Ramaswami Mohandoss

ISBN:978-0-9989746-2-0
(Orange Window Publishing)

DEDICATION

To my sister, Vidhya.

BY THE SAME AUTHOR

In Few Words
What is the P vs NP problem?

ACKNOWLEDGEMENTS

I had access to a lot of great content in books, websites, and YouTube videos. I would like to sincerely thank all the authors and content creators who helped me gain clarity on the various topics related to math and science through their work.

Books

Relativity by Albert Einstein
For the Love of Physics by Walter Lewin

YouTube Channels

Khan Academy
Tyler Dewitt
3Blue1Brown, by Grant Sanderson
Lectures by Walter Lewin. They will make you ♥ Physics.

Websites

www.wikipedia.org
www.britannica.com

CONTENTS

1. INTRODUCTION

Albert Einstein derived this famous *mass-energy equation,* i.e., $e=mc^2$, in one of his published papers in 1905. The English-translated version of that paper is only four pages long. Yet, fully grasping the idea will require fluency in many different concepts (maths and physics). While Einstein formulated most of the ideas himself (Special Theory of Relativity), he did work on the ideas developed by his predecessors, Isaac Newton and James Clerk Maxwell, to name a few. We could derive the mass-energy equation in different ways, and a quick search on YouTube would show all of them. However, our pursuit would be to derive it the way Albert Einstein did in his 1905 paper.

While it may not be feasible to box all physics concepts in one single book, I also believe it is not impossible to detail every concept relevant to understanding $e=mc^2$. In the following pages, we shall build intuition and understand all the basics in detail to become fluent in deriving and explaining this equation. From the standpoint of understanding this equation fully, we shall leave no stone unturned.

Besides the Introduction and Appendix, this book has four sections. In the first section, we shall understand the first fundamental force, i.e., gravitational force. Here, we shall cover classical mechanics that is also called Newtonian physics. In the second section, we shall focus on the second fundamental force, i.e., electromagnetism. We shall start with the basics of electricity and magnetism and derive all of Maxwell's equations. In the third section, we shall focus on the concept of spacetime and Einstein's Special Theory of

Relativity. Spacetime is the paradigm shift formulated by Einstein to resolve a fundamental conflict (more on this later) that existed back then. With all this knowledge under the belt, we will derive the equation with ease in Section 4.

A firm understanding of all the concepts that lead to $e = mc^2$ cannot happen without the help of mathematics. There is no reason to panic here as we will cover all the required math in detail too. We will also be deriving every other equation that is relevant to our understanding of the mass-energy equation. We will cover all the math fundamentals as part of the Appendix section, and you do not have to know all of them before you begin. We shall use the lazy loading technique that I've borrowed from Software Engineering. *Lazy loading* simply means we will learn the basics when we realize its necessity. Some chapters have prerequisites, and we can wait to go through the math fundamentals until we reach those chapters.

In summary, this book assumes minimal mathematical background from the reader's standpoint. If you are familiar with basic Euclidean geometry ($sin\theta$, $cos\theta$ *and* $tan\theta$), you should be good to begin. So, let us get started.

2. SIGHT & LIGHT

2.1 How do we perceive?

In the equation $e = mc^2$, the letter e refers to energy, m relates to mass, and c refers to light speed. Many of you might have a question: What is the speed of light doing in an equation related to mass and energy? We shall start our journey precisely from this point. To decipher the relevance of c in this equation, let us understand the science behind our sight. How do we see an object? We can see an object through the light rays that reflect off the object when they reach the retina of our eyes. In other words, for us to perceive an object, the light rays will have to reach our retina after hitting that object. For light to reach our retina, light has to move. When something moves, it does so at a particular speed.

For a long time, up until the seventeenth century, everyone believed that the speed of light was infinite and immeasurable. That perception changed around 1676 when a Danish astronomer Ole Roemer observed an aberration. Ole Roemer studied Jupiter and its moons and noticed that the eclipses formed by Jupiter's moons were happening sooner than the predicted time when Earth was closer to Jupiter. These eclipses took more time when Earth was farther away from Jupiter. He reasoned the speed of light to be the cause of this phenomenon. When Earth is farther away from Jupiter, light has to travel a much larger distance before it reaches our eyes. Ole Roemer's discovery was a significant milestone that changed the perception of light forever. For the first time, it was becoming evident that light's speed was not infinite but measurable.

Since Ole Roemer's discovery, many scientists and astronomers have attempted to quantify the speed of light. In 1848, Hippolyte Fizeau determined the speed of light to be ~315,000 km/s in a non-astronomical way, i.e., in a laboratory setup. Today, using all modern instruments, the speed of light c is measured to be 299,792,458 meters per second i.e., 186,282 miles per second.

Let us wrap up this chapter with a thought experiment. Let us imagine a setup as shown in figure 2.1.1, where we have two people A and B, separated by distance x. Both A and B are gazing at the sky in the same direction. What would happen if a new star emerged in this line of sight? Since light travels at a finite speed and the star has just appeared, person A who is closer to the star would perceive the star ahead of person B, correct? In other words, is there a point in time, though brief, when the star appears to person A but not to person B? I'll let you ponder this until we discuss this in detail in the later sections.

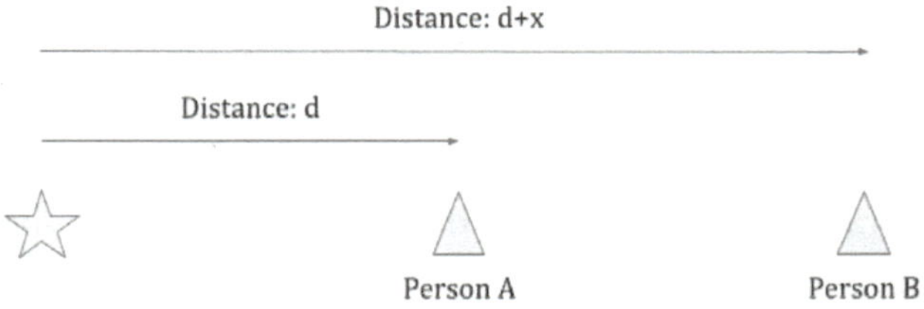

Figure 2.1.1

2.2 Why do objects move?

When man started to see and perceive things around him, what is likely to have intrigued him first was the movement of objects from one place to another. We all must have heard of the story of Sir Isaac Newton watching an apple fall from the tree and how that led to his discovery of gravity. Per Walter Lewin, in his book *For the Love of Physics,* Newton was intrigued why the apple always took a path perpendicular to the earth's surface. Let's put ourselves in Newton's shoes and try to explain the fall of the apple from the tree. I'm well aware that those are large shoes to fill but let us give it a try.

Before we try to explain gravity, let us start with *motion* in general. As shown in figure 2.2.1, we say an object B is moving when it travels with a velocity v relative to an object A at rest. At rest, the velocity of object A is 0. There is a subtle, yet crucial aspect related to motion, and we must understand that nuance before proceeding. We can spot a flying bird, i.e., it is easy to perceive motion. However, feeling motion is not as straightforward. Galileo articulated this aspect of movement in 1632 through the example of a ship traveling at a constant velocity on a smooth sea. He stated that as long as the ocean remains calm and the ship does not rock, any observer below the deck would not be able to tell whether the ship was moving or stationary.

When A and B are the same objects, i.e., when A and B have the same mass *m* and have the same shape, density and volume, the velocity should add something to object B, correct? What is that? That effect that velocity adds to an object is called *momentum* and is quantified as *mass* times the *velocity*. In other words, an object gains momentum by moving and is referred to by the letter *p*.

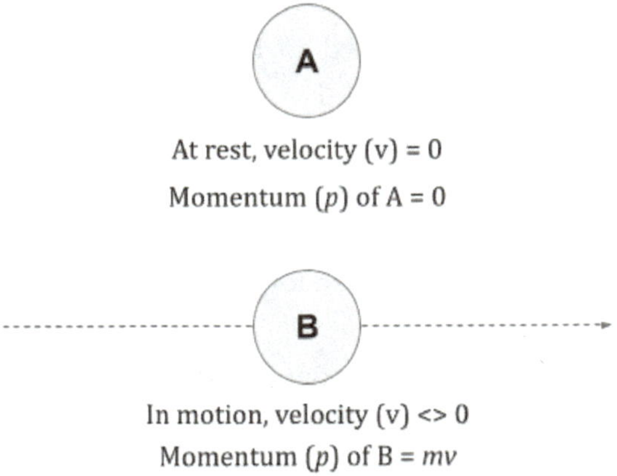

Figure 2.2.1

Now, what could cause object A to change its behavior and start moving? We know that an external action in the form of a push or a pull would accomplish that. In other words, a push or pull would alter some property of object A for it to move and gain momentum. We haven't defined what that property is yet, but let us call it *X*. If we do not alter *X*, object A should remain at rest. Now, what about object B that is already in motion? Here again, if that *X* in B remains unaltered, we would also expect object B to not change its behavior and continue to stay in motion with the same momentum. We can now define Newton's first law of motion: *Until there is an external intervention, an object at rest will remain at rest, and an object in motion will continue to be in motion.*

Newton defined *X* as *inertia.* Inertia is the resistance of any physical object to any change in its velocity. Newton described inertia as the most fundamental property of objects in motion. For this reason, we also refer to Newton's first law of motion as the *law of inertia.* When inertia is undisturbed, we will not be able to feel movement. When Galileo said we could not differentiate between a moving ship and a stationary ship, he referred to undisturbed inertia.

When an external action like a push or pull alters the inertia of the physical object, it does so by exerting an influence on the object. That influence is called *force.* The change in inertia would result in a difference in the object's momentum. Let us consider an object A of mass *m* traveling with a velocity *v.* We know the momentum of object A is *mv.* What would happen when we apply a force F on A for a brief moment (say one second), as mentioned in figure 2.2.2? A's momentum would change to *mv'* and it would continue to move with a velocity of *v'*. What if we apply the same dose of force for another second in the same direction? Object A would now have additional momentum and start moving with an even higher velocity. Let us call the newer momentum *mv"* and the new velocity as *v"*. While we know that $mv'' > mv' > mv$, the difference *mv"-mv'* would be higher than *mv'-mv* because the initial velocity before applying Force *F* (for the same duration of 1 second) was different in both the occasions, i.e., *v* and *v'*.

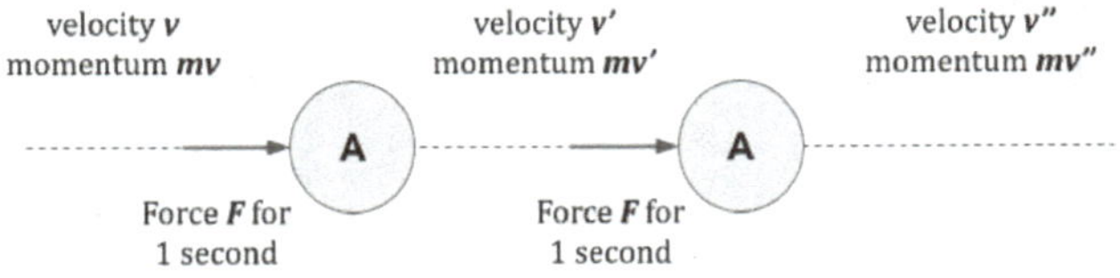

Figure 2.2.2

What if we apply the same force continuously for two seconds, as shown in figure 2.2.3? The object's velocity would increase to *v''*, and the object would move with a momentum of *mv''* directly.

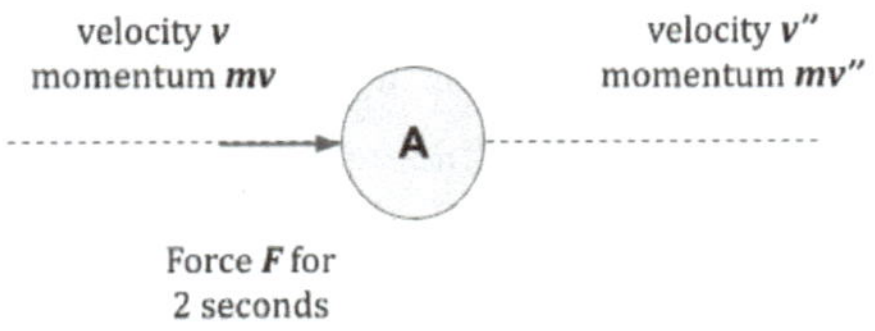

Figure 2.2.3

Through the above experiment, we could see that force alters the inertia of an object, and continuous application of force would continue to modify the inertia at a specific rate. How could we quantify that rate of change of inertia? Since the change in inertia would directly impact the object's momentum, we could infer that through the rate of change of the object's momentum. In other words, we could relate force with the rate of change of momentum as shown below:

$$F \propto \triangle p$$

In the above equation, $\triangle p$ is the rate of change of momentum after applying the force on the object for *t* seconds. If *v* is the initial velocity of the object and *v''* is the velocity of the object after applying force *F*, we could write the above equation as below:

$$F \propto \frac{mv'' - mv}{t}$$

$$F \propto m\left(\frac{v'' - v}{t}\right)$$

$$F \propto ma$$

In the above equation, *a* is the rate of change of velocity, also called the *acceleration* of the body. By introducing a constant *k*, we could rewrite the above equation as below:

$$F = kma$$

If we consider *k* to be equal to 1, then *F* equals *ma*. Can *k* be any other number? Sure it can, but remember that Newton conceived the idea of force, and he chose to quantify one unit of force this way.

$$F = ma \quad (2.2.1)$$

In other words, *F* denotes the force required to make an object of mass 1kg (SI unit for mass is kilogram) acquire an acceleration of 1 ms^{-2}. The amount of force needed to accomplish this is *1 Newton*. We have just defined Newton's second law of motion: *The force acting on an object equals its mass times the acceleration.* The unit for Force is *Newtons,* and we can see from the above equation that:

$$1 \text{ Newton} = 1\text{kg ms}^{-2}$$

How does a sprinter take off in a race? How does a professional swimmer launch herself into the water to begin her race? If you observe their feet closely, be it the swimmer or the sprinter, in both the occasions, to leap forward, they will apply a good amount of force in the opposite direction. To be precise, the sprinter would apply the force on the starter block, and the swimmer would apply the force on the launchpad. Here comes an essential property of force: Forces always occur in pairs. To leap forward, one has to push backward. We are ready to define Newton's third law of motion: *Every action has an equal and an opposite reaction.*

Now that we have a handle on Newton's three laws of motion, it would be good to derive some equations and have them in our back pocket. We shall derive three equations commonly called the *three kinematic equations of motion from Newton's laws of motion. Kinematics* is a branch of physics that deals with the motion of objects. These equations help describe the motion of a body in terms of its three kinematic variables: velocity (v), position (s), and time(t). These equations shall come in handy as we decipher more concepts in the subsequent chapters.

Let us consider an object A of mass m under a constant influence of force F. According to Newton's second law, A would move with a constant acceleration a. We have depicted this in the following graph (figure 2.2.4), where we have plotted the object's velocity against time. Between 0 and t, *v* is the object's velocity at time 0, and *v'* is the object's velocity at time *t*.

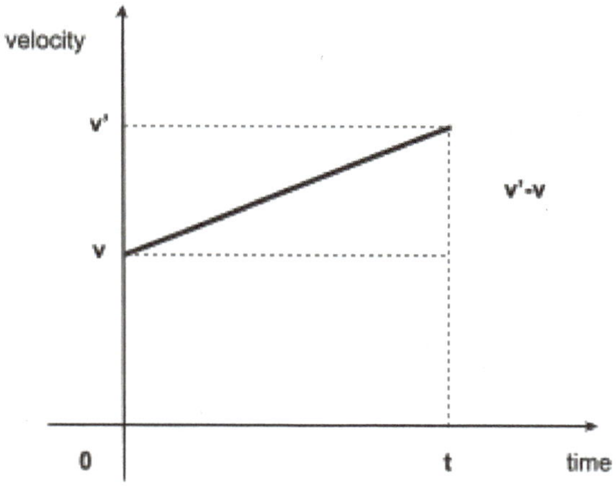

Figure 2.2.4

We can derive the first kinematic equation of motion from Newton's second law of motion. We can derive the acceleration *(a)* experienced by object A as the following:

$$a = \frac{v' - v}{t}$$

$$v' = v + at \qquad (2.2.2)$$

In the second kinematic equation of motion, we shall derive the distance covered by the object in time *t*. Before we calculate that, let us zoom in to figure 2.2.4. Each rectangular block in figure 2.2.5 shows the object's velocity (between *v and v')* in one unit of time. So, the distance covered by the object in *t* units of time is a sum of all these values. In other words, the distance covered (*s*) equals the area of the trapezium formed by the edges *(0,0)*, *(t,0)*, *(t,v')*, and *(0,v)*.

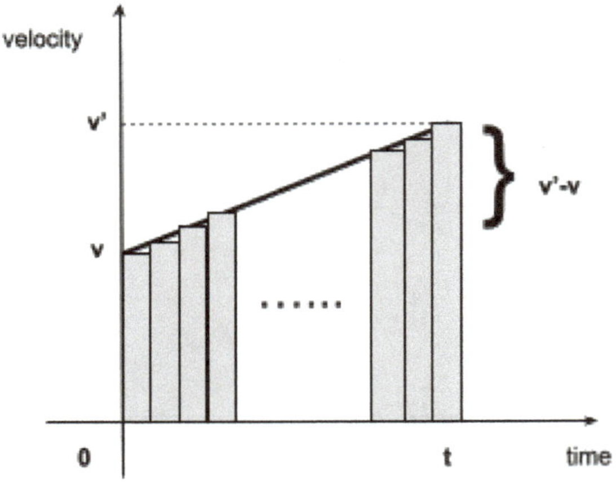

Figure 2.2.5

We are now ready to derive the second kinematic equation of motion.

s = Area of the trapezium = Area of the rectangle + Area of the triangle

$$s = vt + \frac{1}{2}(v' - v)t$$

Substituting for *v'* from equation 2.2.2

$$s = vt + \frac{1}{2}(v + at - v)t$$

$$s = vt + \frac{1}{2}at^2 \qquad (2.2.3)$$

The third kinematic equation of motion is a variation of the second one. Here too, we will be deriving for the distance covered (*s*), but we shall do that without the help of time(*t*). Instead of splitting the trapezium into a rectangle and a triangle, we will derive the area of the trapezium directly using the below formula.

$$s = Area\ of\ trapezium$$

$$s = \frac{1}{2}t(v + v')$$

Substituting for *t* from equation 2.2.2

$$s = \frac{1}{2}\left(\frac{v' - v}{a}\right)(v' + v)$$

$$2as = v'^2 - v^2 \qquad (2.2.4)$$

We have derived all the three kinematic equations of motion based on Newton's laws of motion. But, what about the original question? What made the apple fall? How was Newton able to crack the puzzle of gravity? Read on.

While Newton was able to explain the fundamentals of motion through his three laws, these were not sufficient to explain the mysterious force of gravity. As Newton tried to make sense of gravity, he was equally puzzled by the behavior of the moon and the other planets. What made the moon orbit the earth, and what made the earth and the other planets orbit the sun? These were other questions that bothered him. That's when Newton performed a brilliant thought experiment, now called *Newton's cannon experiment.* The experiment goes like this. Let us imagine a cannon placed on a tall hill. What would happen if we fire the cannonball in the direction parallel to the surface of the earth with a velocity *v*? Based on Newton's laws of motion and the force of gravity, we expect this cannonball to travel a certain distance and eventually yield to the force of gravity and fall at a distance from the hill. What if we fire another cannonball with a higher velocity *v*? We would think the ball would exhibit the same behavior as before and would fall at a different point from the hill. These are marked A and B in figure 2.2.6.

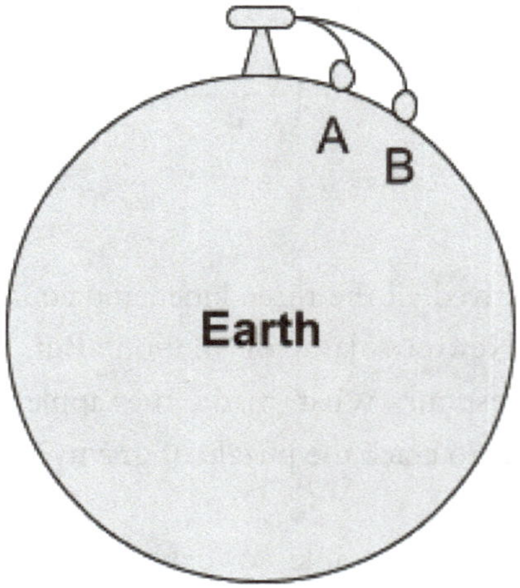

Figure 2.2.6

Let us take the thought experiment a little further. What would be the possible outcome when the cannon fires with an extremely high velocity? To be precise, a velocity so high that the cannonball travels a long distance before yielding to the force of gravity. Also, when the cannonball begins to fall, the earth's curvature significantly increases the distance between the cannonball and the earth's surface (compared to the hill's height). By the time the cannonball travels to cover that extra distance, the curvature recreates that same distance. Newton imagined such a cannonball to fall eternally towards the earth but never touch the surface of the planet earth. In other words, such a cannonball would continuously orbit the earth. We have depicted this in figures 2.2.7a and 2.2.7b. At this very moment, Newton understood the explanation behind gravity and answered his questions on how planets and stars behaved.

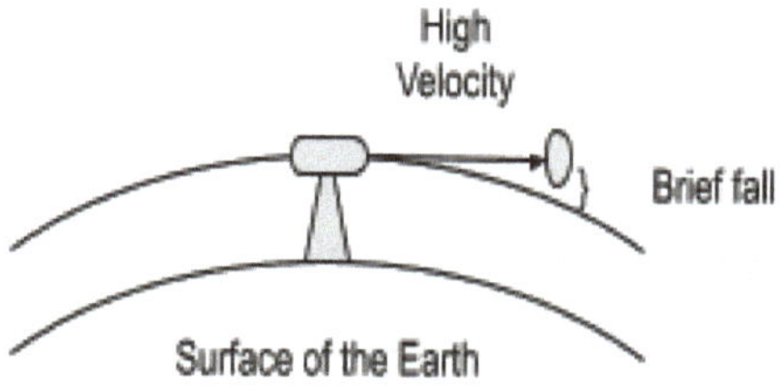

Figure 2.2.7a

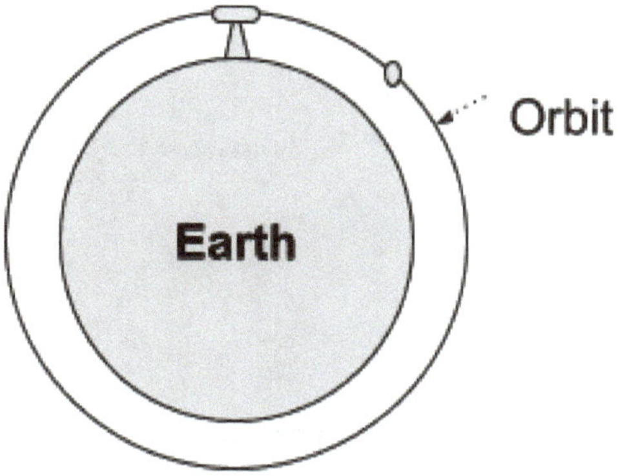

Figure 2.2.7b

Why does the moon revolve around the earth? Newton reasoned that the earth attracts the moon.

Why does the moon not fall on the earth? Newton reasoned the attraction to fade as we move away from the earth. The moon is indeed trying to fall, and the earth's curvature is causing it to eternally orbit. In essence, the moon is behaving like a cannonball.

Newton formulated the *Universal Law of Gravitation.*

According to the law, *every particle in this universe attracts every other particle with force directly proportional to the product of their masses and inversely proportional to the square of the distances between their centers.* We have shown this in figure 2.2.8, where the mass of object A is greater than that of object B, i.e., $M_A > M_B$.

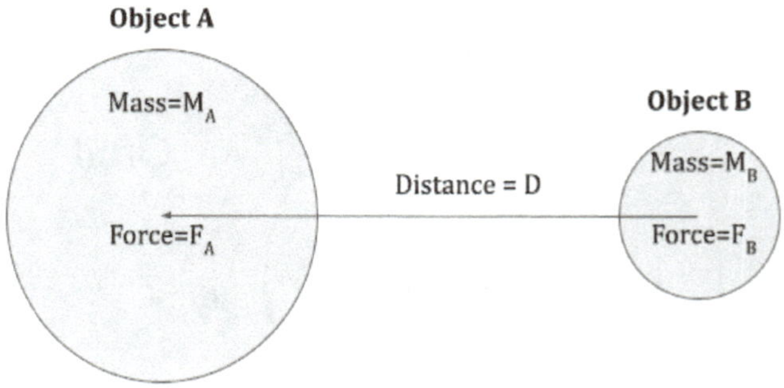

Figure 2.2.8

$$F_A \propto \frac{M_A M_B}{D^2} \; and \; F_B \propto \frac{M_A M_B}{D^2}$$

Or

$$F_A = F_B = G\frac{M_A M_B}{D^2} \qquad (2.2.5)$$

In the above equation, *G* is the *Gravitational constant.*

According to the Universal law of Gravitation, the sun being the heaviest star attracts all the planets. Every planet based on its mass stays at an appropriate distance (based on the above equation) and orbits the sun. When an apple falls towards the earth, it is not only that the earth is attracting the apple, but the apple is also attracting the earth with the same force. Since the apple's mass is negligible compared to the earth's, the apple ends up accelerating towards the earth. In other words, the apple falls on the surface of the planet Earth.

When Newton formulated the law of gravitation, he hadn't calculated the value of the gravitational constant *G*. Around 111 years later, in 1798, British scientist Henry Cavendish tested this law in a laboratory. This experiment is now called the Cavendish experiment. Cavendish derived the value of G to be the following:

$$G = 6.67430 \times 10^{-11}\ m^3kg^{-1}s^{-2}$$

2.3 How to make an object move?

In the previous chapter, we touched upon applying force to alter the inertia of an object. However, we did not get into the details of the source of that force, which would precisely be our focus here. As shown in figure 2.3.1, let us take the context of the snooker table. The table has a ball (A) at rest, and our pursuit is to lead it to one of the corner pockets.

For the ball to reach the pocket, the ball would need to move, and for the ball to move, we would need to alter its inertia. How would we move the ball? We know that we could use the cue stick. Let us closely examine what we could do with the cue stick.

To pocket the ball using the stick, we would need to place the cue stick in line with the path to the pocket and sharply move the cue stick forward such that it applies a force on the ball the moment it hits it. If the cue stick moved with the proper velocity and the right direction, the chances are that we would have pocketed the ball. What we just defined is the *work* done by the cue stick to move the ball. In other words, the cue stick will have to perform work to create a force on the ball or any object.

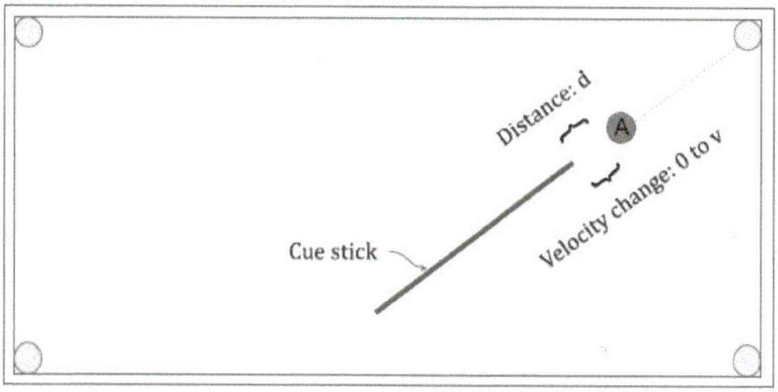

Figure 2.3.1

Let us try to quantify the work performed by the cue stick. What happened to the cue stick was that it had to reach a certain velocity v after covering distance d before touching the ball. Before completing this work, the cue stick was at rest, i.e., the initial velocity was zero. If the velocity of the stick increased from 0 to v after covering a distance of d units, it must have undergone acceleration. If M_s is the mass of the stick and A_s is the acceleration of the stick, then the force experienced by the cue stick is M_sA_s.

In physics, we define *work* as the force (experienced by an object) times the displacement of that object. In our scenario, work performed by the stick (denoted by W_s) equals the force experienced by the stick (M_sA_s) times the displacement, d.

$$W_s = M_sA_sd \qquad (2.3.1)$$

Given that the cue stick has acquired a velocity of v from 0, from the third kinematic equation of motion, we know that,

$$d = \frac{v^2 - 0^2}{2A_s} = \frac{v^2}{2A_s}$$

Substituting the above for *d* in equation 2.3.1, we get:

$$W_s = \frac{1}{2} M_s v^2 \qquad (2.3.2)$$

If you observe, the work done by the stick as calculated through the above formula has the following unit: kg m^2 s^{-2}. However, the unit for work is *Joule*, which is also the unit for *energy*. So, 1 Joule equals kg m^2 s^{-2}. The mechanical movement of the cue stick has helped it acquire a certain amount of *mechanical energy,* called the *kinetic energy* that was used to apply the force on the ball. The adjective *kinetic* means *relating to or resulting from motion,* and the motion of the cue stick helped the stick acquire the energy. The movement of the stick was made possible through the work done by the hands of the player. In other words, the energy acquired by the cue stick was transferred through the work performed by the player's hands.

To summarize, the work done by the hand resulted in the stick's movement, and the work done by the stick resulted in the ball's movement. Thus, work and energy are one and the same and two sides of the same coin. We transfer energy from one object to the other through work, and for an object to perform work, it would need energy. We have articulated this in figure 2.3.2.

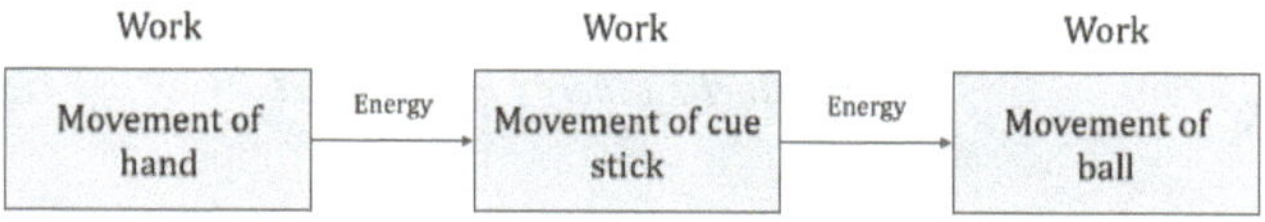

Figure 2.3.2

So far, we saw how we could transfer energy from one object to another through motion, and we called it kinetic energy. Now let us consider the scenario of an object that is at rest. In figure 2.3.3, we have a string holding a ball of mass *M*. The string holds the ball at a distance (or height) *h* from the ground. Since it is above the ground, it is under the influence of gravity and let us call it *g*. What would happen if we snapped the string above the ball? The ball would yield to the effect of gravity and hit the ground, correct? In other words, just like the cue stick, right when the ball hits the ground, it would have acquired energy through its work (movement towards the surface). Let us try to quantify that Work (*W*).

$$W = \textit{Force experienced by the ball} \times \textit{distance traveled by the ball}$$

$$W = Mgh \qquad (2.3.3)$$

The ball will perform the work only when we snap the string. Until the string snaps, we could say that the ball can complete the work of *Mgh Joules,* correct? What we have just defined is the *potential energy* of an object. The Potential Energy of an object is its capacity to do work while it is at rest. The total mechanical energy is, therefore, the sum of its kinetic energy and potential energy.

Total Mechanical Energy of an object = Potential Energy of the object + Kinetic Energy of the object

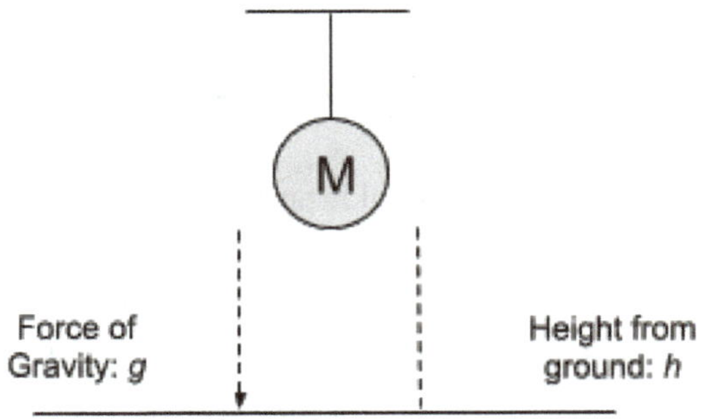

Figure 2.3.3

Let us focus on kinetic energy again. Through the examples of the cue stick, we have learned that the transfer of energy from one object to another happens through mechanical work, i.e., motion. The ball moved because the cue stick moved, and the cue stick moved because the player's hand moved. Now, you might have this question: What moved the player's hand in the first place? In other words, how did the player acquire that initial energy to push his hand? While beyond the scope of this book, mechanical energy is not the only kind of energy. There are other kinds of energies too—thermal, chemical, nuclear and others. In the scenario of the snooker player, the chemical energy the player would have consumed from the food eventually gets transformed into mechanical energy. We have articulated the transformation of energy in figure 2.3.4.

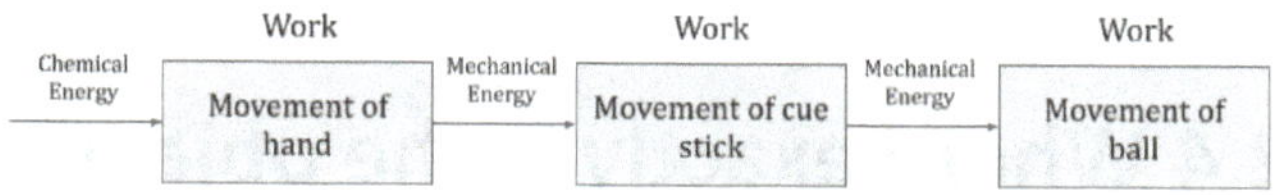

Figure 2.3.4

We are now ready to learn an important law of work and energy: *The law of conservation of energy*. The law of conservation of energy states that the total energy of an isolated system remains constant. Energy can neither be created nor destroyed; instead, it can only be transformed or transferred from one form to another.

2.4 Where exactly is the object?

Newton's laws are no doubt a milestone in the history of physics. Through those three laws, Newton had almost succeeded in explaining the behavior of objects in motion thoroughly. Why I say "almost" will soon be apparent. While we have derived Newton's laws and the three kinematic equations of motion, there is one aspect that we are yet to detail out formally. In all the examples shared so far, be it the movement of the cue stick on a snooker table or the static hanging ball, we measured the object's position, both at rest and motion, in a fixed frame of reference. A frame of reference is a system of coordinates. For instance, figure 2.4.1 captures the displacement of the cue stick as a shift in position from *(x', y', z')* to *(x", y", z")* in a specific system of coordinates. That system of coordinates is a three-dimensional space with three axes perpendicular to one another.

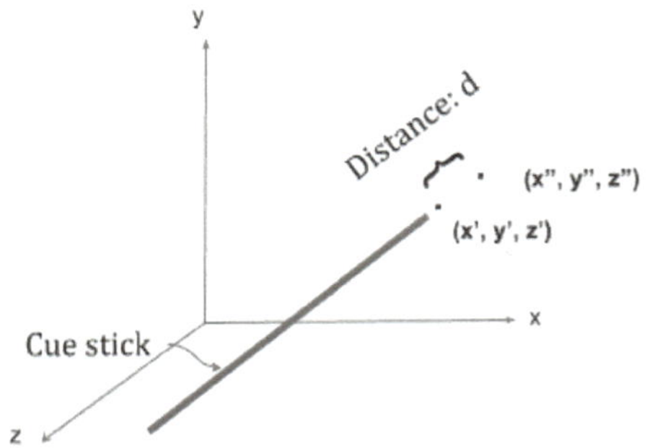

Figure 2.4.1

Why do we have to make the frame of reference explicit? Let us understand that through a thought experiment. Let us consider the following scenario as shown in figure 2.4.2: We have two people, A and B, a moving container, and a ball (b). Person A is standing in the container and is holding the ball in her hands. The container is moving with a constant velocity v. Person B is standing on the stationary pavement and is watching the container move. The container appears transparent from the outside but opaque from the inside. In other words, Person B can see Person A and the ball b, but Person A cannot see anything outside the container. Since the container, person A and ball b are moving at a constant velocity, as per Newton's first law, the inertia experienced by all three would be the same. In other words, since Person A cannot see anything outside of the coach, she would perceive everything around her to be at rest. Person A would believe her frame of reference to be stationary. Let us call it K. In K, the ball b is not moving and resting. However, from the perspective of Person B, the ball is moving with the velocity of the container, i.e., v. In other words, relative to the frame of reference of person B (let us call it K'), the ball and the frame of reference K is in motion with a velocity v. To summarize, between K and K', K' is a fixed frame of reference that determines the ball to be in motion.

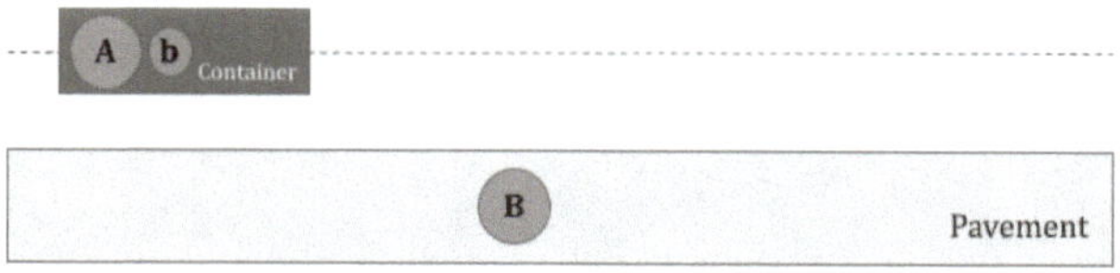

Figure 2.4.2

In the above example, though Person B might think that his frame of reference (*K'*) is stationary, the pavement is built on the earth, which revolves around the sun. So, the pavement is moving relative to the sun, and *K'* isn't fixed either. Is the position of the sun fixed? We don't have evidence of that either. Through the *Big Bang theory*, the belief is that the universe is constantly expanding. We have come to realize something crucial—we are yet to discover a fixed frame of reference. In other words, there is no absolute space to which we could pinpoint an object. The kinematic laws of motion derived from Newton's laws are always relative to a specific frame of reference, hence the need to explicitly call it out. That particular frame of reference (aka system of coordinates) that Newton's laws hold relative to is called the Galilean system of coordinates. We call it Galilean because Galileo realized the need for multiple frames of reference in 1632.

When there is no absolute space, or in other words, when there is no fixed frame of reference, it becomes inevitable for us to devise a method to transform from one frame of reference to another. This is what we will tackle in the next chapter.

2.5 The Galilean transformation

In the thought experiment that we had last chapter, Person B, who is at rest relative to the container, continues to observe the motion of Person A. Person A remains blissfully unaware of her state of motion and continues to believe that she is at rest too. Let us extend that thought experiment.

In a while, Person B realizes that there is a restricted military zone in the container's path. We have shown this in figure 2.5.1. If the container continues to follow the path line, it will soon enter the restricted zone. If that happens, the army could spot the container and might retaliate by shooting the container, proving dangerous to person A.

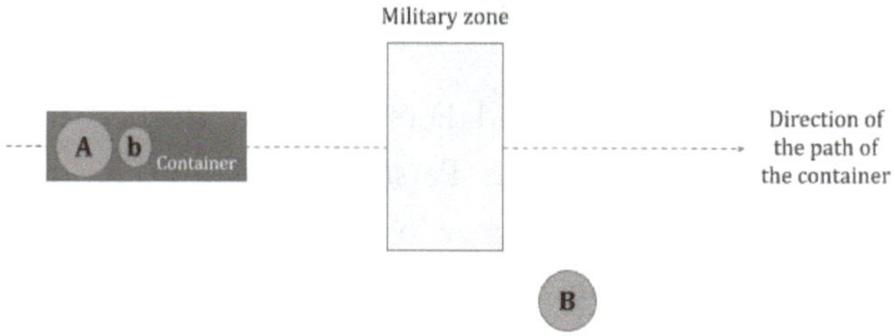

Figure 2.5.1

In the extended thought experiment, Person A and B can communicate with each other through a phone. In other words, there is a phone in the container and the hands of Person B. As a good Samaritan, considering the impending danger for person A, Person B picks up his phone and reaches out to Person A.

Person B: Hello!

Person A: Hello!

Person B: Sorry to trouble you. I wanted to reach you to convey something that you might want to hear.

Person A: Sure, go ahead.

Person B: There is a restricted military zone in the direction of your motion. Entering that might not be a good idea.

Person A: Wait a sec, I don't think I'm moving. (Remember, the frame of reference for person A is *K*, and everything in the container has the same level of inertia. Also, nothing outside the container is visible to Person A.)

Person B: Well, I have evidence that you are moving. You should either change your direction or stop to avoid entering the restricted zone.

Person A: I understand your intent to help. But I think you are mistaken. Thanks for calling.

Person A hangs up, and Person B is disturbed that he is unable to help. Though Person B was able to detect the movement of the container in his frame of reference, *K'*, he was unable to convince Person A. They did not have a common ground between their frames of reference (*K* and *K'*). After a while, Person B receives a phone call. The caller's number is familiar to Person B, who identifies the number as person A. He quickly picks up the phone and is happy to hear the voice of person A on the other side. The conversation is a little different this time.

Person A: Hello, there!

Person B: Hi! I hope you are doing okay.

Person A: Yes. Somebody did attack my car (container) from the outside, but fortunately, nothing happened to me. I now think it could be the army from the military zone that you spoke about some time back.
Person B: Yes, considering the situation, that is quite possible. But I'm glad that you have survived.
Person A: I need a favor from you.
Person B: Yes, tell me.
Person A: None of what is happening around me gives me any evidence that I'm moving. (Remember, the container is closed, and *A's* frame of reference is *K*.) However, I cannot ignore your advice, either. Could you please tell me if there is any other impending danger?
Person B: Let me check.

Person B places A on hold briefly and checks his map. To his shock, he realizes that the container is indeed approaching another military zone (figure 2.5.2). In his frame of reference, *K'*, person B realizes the velocity of the container to be 60 kilometers per hour, and the container is approximately 10 km away from the next military zone. The second military zone is 5 kilometers wide.

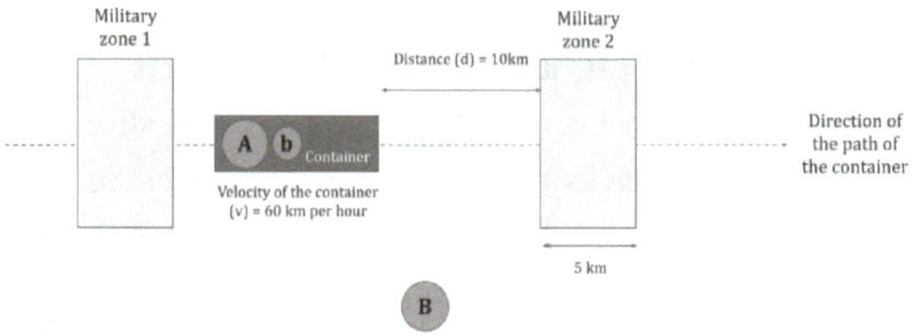

Figure 2.5.2

Taking learning from his earlier conversation with person A, person B tries something different this time.

Person B: Hey, you there?
Person A: Yes.
Person B: Unfortunately, you are indeed approaching another military zone in 10 kilometers. I'm also aware of your situation that is making you believe that you are not moving. So, I'm going to propose something else. Would you be open to that idea?
Person A: Yes, please tell me. I don't think I have an option.
Person B: Do you have a clock with you?
Person A: Yes, I do have one.
Person B: Great. Here is my suggestion. Please reset your clock and wait for the tenth minute. On the tenth minute, switch off your lights and stay with your lights off for another 15 mins. At the end of the fifteenth minute, you could bring back your lights on. This act should help you sail through the military zone inconspicuously and minimize your chance of getting noticed.
Person A: Sure, thanks. I'll follow your instructions.
Person B: Good Luck!

Person B drops off, and after 20 mins, he gets a call back from Person A, who confirms that he is still alive and that there were no attacks this time on the container. So, how did both Person A and Person B manage to coordinate and successfully navigate the second military zone? Despite the lack of common ground between their frames of reference (*K* and *K'*), they managed to find common ground through the clock. If we cannot pinpoint the absolute position of an object, could we identify the presence of an object's location at a particular point in time? With time as the additional dimension, and with the help of a *synchronized clock,* Person B was precisely able to advise Person A. What person B just did was to transform an event from his frame of reference *K'* (denoted by <*x', y', z'*> and time *t'*) to an event in the frame of reference *K* (denoted by <*x,y,z*> and time *t*). Technically, this is called the *Galilean transformation.* We can use this method to transform events between the coordinates of two reference frames (*K* and *K'*) which differ only by constant relative motion. We've represented this pictorially in figure 2.5.3.

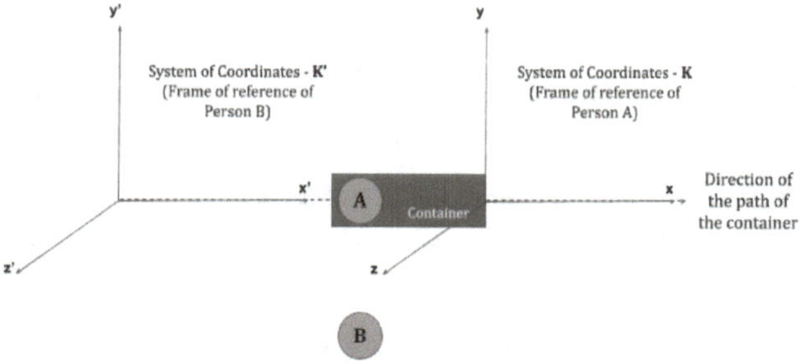

Figure 2.5.3

As you might observe through the above figure, the dimensions of two frames of reference, *K* and *K'*, are related through the following equations. Here, *K* travels with a velocity of *v* in a specific direction relative to *K'*.

$$x' = x + vt'$$
$$y' = y$$
$$z' = z$$
$$t' = t$$

2.6 The illusion of simultaneity

Let us summarize our journey so far. A couple of chapters ago, our thought experiment had led us to a significant limitation: the absence of *absolute space*. We then learned how to overcome that limitation using a synchronized clock. Though Person A and Person B could not align and agree on their relative movement, with the help of a synchronized clock, they could precisely align on the timing of a specific event. Then with the aid of the Galilean transformation, we found a way to transform from one frame of reference to another. Remember, in our pursuit to overcome the limitation stated earlier; we had made a crucial assumption. The assumption is that, with the help of synchronized clocks, Person A and B should both perceive an event and time that event in the same manner. In other words, the assumption is that two people from different frames of reference can align on the absolute time of an event. Is that assumption valid? Let us explore that in this chapter.

A clock is merely a mechanical device that helps us keep track of time. What exactly is time? Time is the progression of an event from the future to the present and then into the past. An event that we perceive now is the present. Before we perceive it, the event falls in the future, and once we perceive it, it falls in the past. A clock merely helps us index the present moment with a unique number, which we then use to record and remember the correct sequence of events. In our thought experiment, when person B advised person A to turn off the light in the tenth minute, he had assumed two things.

1) The watch held by person A behaves the same way as the watch held by person B.
2) More importantly, person B has assumed the present moment that he perceives is also perceived simultaneously by person A. In other words, when person B perceives the container to touch the military zone, he assumes person A would also perceive the same event at the very moment had person A been able to look outside the container. In other words, person B believed they both perceive an event simultaneously.

The second assumption is harmless only when the container moves at a low speed, relative to the speed of light. What would happen when the container (or the frame of reference K) is moving fast? Let us understand the repercussions through another thought experiment, which I've borrowed from the writings of Albert Einstein himself.

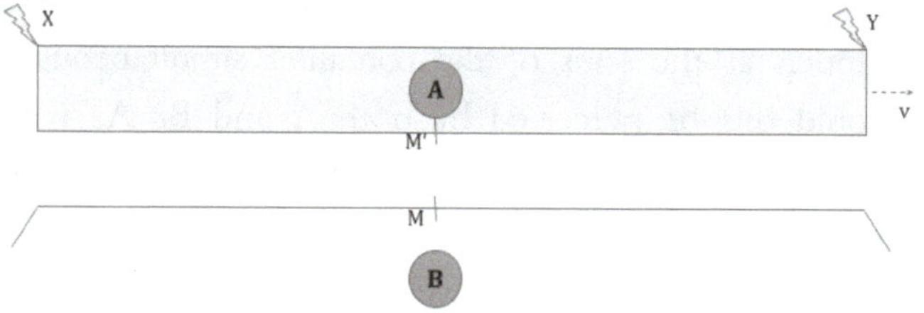

Figure 2.6.1

Our thought experiment here will involve a setup similar to our earlier one but with some changes. In this thought experiment, we shall consider the following:

- The container is much longer and is traveling with a high-velocity *v*.
- The container is now not opaque from inside. It has windows on all four sides through which person A could perceive what is going on outside.
- All the air is fully sucked out on top of the pavement and inside the container. In other words, we fill the space with a vacuum. (The reason for this assumption will soon be apparent.)
- Person A is standing right in the middle of the container, and Person B is standing on the pavement.

The frame of reference of person B continues to be *K'* and the frame of reference of person A continues to be *K*. Right at the moment when the container passes through the pavement and when Person A comes in alignment with person B, let us imagine this scenario: Two lightning strikes happen at the ends of the container simultaneously. How would this be perceived by both A and B? As you might observe in figure 2.6.1, person B is right at the midpoint (marked *M*) of the length of the container. The two lightning strikes would need to travel the same distance from the lightning spots (*X* and *Y*) to reach person B. So, person B would perceive the lightning to have struck simultaneously.

How about person A? Let us dissect that. From figure 2.6.1, you might observe that person A is standing in the middle of the container, and the container now has windows, so the lightning is visible to person A. When the lightning strikes the ends of the container, person A is also equidistant from the lightning spots. But remember that the container is moving at an extremely high velocity toward the right. So, the position of person A, i.e., *M'* as a midpoint to the two flashes of lightning, is only for a very brief moment, as person A would immediately start moving towards the right, that too fast. Right after the lightning strikes, person A will start moving towards *Y* and farther away from *X*. In other words, person A is likely to perceive the lightning that struck at *Y* ahead of the one that struck at *X* because light would have had a shorter distance to travel from *Y*, i.e., $M'Y < M'X$.

This thought experiment has now led us to a startling revelation. Two events perceived to be simultaneous by person B are not perceived the same way by person A, traveling inside a high-speed container. Einstein calls this *Relativity in Simultaneity* in his writing. In other words, there is no *absolute time* either.

You might think of another possibility here. We had already assumed the medium to contain a vacuum. Is it possible that the light is capable of moving with a different velocity inside the moving container? That way, it would cover the distance *XM'* quicker, allowing person A to perceive the two lightning flashes as simultaneous events. Here comes the exciting part. Recent discoveries in physics have shown through evidence that the speed of light in a vacuum is indeed constant and independent of the medium's motion. In other words, the speed of light both inside the container and outside the container would remain the same. In our next section, we shall focus on the journey that led to this revelation.

3. SCIENCE OF LIGHT

3.1 Throwing some light on light

In the opening chapters of the previous section, we did see that interest in the light had sparked way back in 1676 when Ole Roemer measured the speed of light for the first time. Though many physicists explored the behavior of light, the science of light remained a mystery up until 1862. The man who threw light upon light is James Clerk Maxwell. Between 1861 and 1862, Maxwell proposed that light is an electromagnetic phenomenon through a set of mathematical equations, now called Maxwell's equations.

We dedicated the first section to Sir Isaac Newton and the first fundamental force, i.e., gravity. Similarly, we will dedicate this section to Maxwell and his discovery of the second fundamental force, i.e., electromagnetism. Besides helping us understand the science of light, Maxwell's equations and a few other experiments will lead us to an important limitation about time. The limitation is that the idea of *absolute time* is untenable. When you experience this problem for the first time, it's likely to make you feel caught in a checkmate situation. Reaching that situation is a perfect moment for us to understand the genius of Albert Einstein. In the next section, we shall see how Einstein relieves us from this tight corner through his super simple idea of *spacetime*. Before we get there, we need to go through Maxwell's journey first and understand what prompted Maxwell to propose light as an electromagnetic wave. To understand electromagnetism, we would need to understand what electricity and magnetism are. Let us start with a brief history of electricity.

3.2 A brief history of electricity

I have tried this multiple times at home, and this should work for you too. The next time your TV remote control stops working, do not go for a replacement battery immediately. Remove the old battery, rub the same with your hands for a few seconds, put them back, and try using your remote again. You are likely to see it work for a few more hours or even days. In the following few pages, our pursuit would be to understand the science behind this.

People have experienced electricity since civilization. The history of electricity dates back to the ancient Greeks, who observed a rod made of amber (a fossilized resin) to attract dry leaves after rubbing the rod with a cloth. It is also likely that people recognized electricity from lightning. For gravity, we have the story of Newton and the apple. But for electricity, it is unclear what piqued man's curiosity to understand and explain electricity fully.

Similar to how Maxwell managed to explain the science of light (that we shall soon see) after decades of exploration by physicists, the science of electricity too got explained only decades after it was discovered and even commercialized. Benjamin Franklin is one of the many physicists who explored and tried to make sense of electricity in the eighteenth century. Before we see how Franklin explained electricity, let us summarize how people experienced the behavior of electricity in the eighteenth century.

Early in this chapter, we had explained that Greeks were the first to observe the effects of electricity when they observed an amber rod attracting dry leaves after being rubbed against a cloth. That was only part of the story about electricity. The Greeks also observed a glass rod exhibited a different behavior after rubbing it with the same cloth. They learned the difference through the below steps.

- Two amber rods repelled one another after rubbing them with a cloth.
- A glass rod and an amber rod attracted one another after rubbing them with a cloth.

In other words, electricity exhibited two polarizing behaviors, and the first plausible explanation came from Benjamin Franklin.

He proposed treating electricity as a *fluid* that can move from one substance to another. Some substances have more of that fluid, and some have less of it. When the fluid moves from one substance to the other, it changes the state of the two substances. Since the two possible states were polarized (glass rod attracting an amber rod after being rubbed by the cloth), Franklin decided to give complementing names to the two states. He named the two states: *positive and negative.* By giving the states a name, Franklin intended to capture the relative difference in the behavior of the two rods (amber and glass) after being rubbed by a cloth. Apart from this, the words positive and negative do not mean anything.

Franklin could have chosen either way, but he decided to state that the glass rod became positively charged after it was rubbed with a cloth. The previous statement meant the amber rod had to become negatively charged after it was rubbed with a cloth. With this understanding, we could now envision electricity as the flow of negatively charged fluid from one substance to another. The glass rod becomes positively charged after losing some negative charge to the cloth. Similarly, the amber rod would become negatively charged after acquiring some negative charge from the cloth. Though the concept of charge came much later, Franklin's core idea of envisioning electricity as a fluid and the polarity (positive and negative) gained acceptance amongst physicists and scientists.

If you have been observing closely, Franklin's idea of electricity as a flow of negative charge and calling the glass rod to become positively charged implied the presence of another element, i.e., a positive charge in the same substance. That is inevitable because for a substance (glass rod or the amber rod or the cloth) to stay balanced before being rubbed, the negative charge should be kept intact by an equal amount of positive charge. Though Franklin did not have any experimental evidence to back his theory, many scientists built upon Benjamin Franklin's work. One of the crucial milestones was the invention of the light bulb in 1879 by Thomas Edison.

But what exactly are these negative and positive charges inside a substance? Do they even exist? We found answers to these questions at the end of the nineteenth century and the beginning of the twentieth century when we started getting an idea about what was happening inside an atom.

3.3 Inside an atom

Prerequisites: To better understand this chapter's content, I would recommend a refresher on the following topics covered as part of the appendix: *Vectors* & *Alternate Coordinate Systems.*

It is time for another thought experiment. Let us consider a wooden stick and an iron rod. We know that the iron rod is stronger than the wood, i.e., it can hurt more when used to hit somebody. But what makes it strong? If we cut the wooden stick (or the metal rod) into two halves and repeat this process an infinite number of times using an imaginary tiny knife, what will we get at? Or how long can we proceed with this process of cutting them into halves? Such questions confounded physicists and chemists, and for many years, there were no clear answers. It was not until a few hundred years ago that they could reach a consensus through a unifying theory.

The man who pioneered that unifying theory is John Dalton. He proposed that tiny indivisible atoms make matter (wooden sticks, metal rods, etc.). Per this proposal, our pursuit of splitting the metal rod can only progress a finite number of times until we arrive at the atom of iron. Dalton made this proposal based on two fundamental laws that physicists and chemists had generally agreed upon. The first is the *Law of conservation of mass*, and the other is the *Law of constant composition*. Let us briefly look into the two laws.

The law of conservation of mass is born from the ancient Greek idea that the total amount of matter in the universe is a constant. When we add a certain quantity of salt (say 100 grams) to water (say 1000 grams), salt would dissolve in the water to result in 1100 grams of saltwater. That forms the crux of this law, i.e., despite chemical reactions and physical transformations, mass is conserved. Formally, we state the law of conservation of mass as follows: *matter is neither created nor destroyed in a closed, i.e., isolated system.*

Let us come to the second law. We can feel that the iron rod is stronger than the wood, which means there is some property of iron that makes it stronger relative to the wood. That property should remain in the infinitesimal part of the iron that we make out of the rod. In other words, the composition of elements that define iron's property should remain constant, however small we go. This property forms the crux of the law of constant composition. We can now formally state this law: a *pure compound will always have the same proportion of the same elements.*

Through his atomic theory, John Dalton was able to give a plausible explanation for many unanswered questions. Yet, the science of electric force remained unexplained. In 1897, almost two decades after Thomas Alva Edison patented his electric bulb, the science of electric force became clear. That happened through an experiment conducted by Sir Joseph John Thomson (JJ Thomson) in October 1897.

In his experiment (figure 3.3.1), Thomson took a vacuum tube that had two metal plates on the left. The only difference between the two plates is that the second one had a small hole in it. Thomson attached a battery to the two metal plates. As we had mentioned earlier, the science of electricity came much later to discovering and commercializing electricity. When Thomson performed this experiment in 1897, batteries were already in the market. The Italian scientist Alessandro Volta invented electric batteries in 1800.

In the experiment, Thomson attached the negative end of the battery to the first metal plate and the positive end of the battery to the second metal plate. If substances contain positive and negative charges (as hypothesized by Franklin) through this setup, the negative charge will get pushed to the rightmost end of the first plate. Same way, the positive charge would get pushed to the leftmost end of the second plate. The placement of the two plates will attract the negative charge from the first plate towards the two, and some of them should escape through the small hole.

Thomson expected to see some change in the right-hand side of the vacuum tube and what he saw was a glow to appear on the right end of the vacuum tube in line with the hole on the second metal plate.

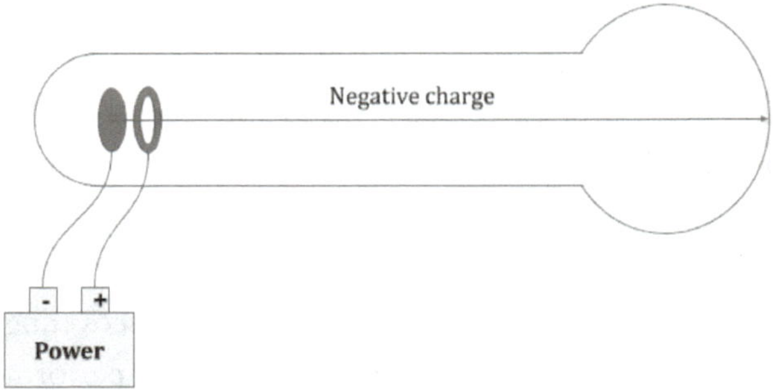

Figure 3.3.1

Thomson reasoned that what appeared as a glow was the negative charge that flowed from the first metal plate. In Franklin's words, this was the electric fluid. Thomson wanted to be sure that this indeed was the case. To verify that, he performed a few validations. One of them involved placing two other metal plates in the middle of the tube and attaching them to a battery. He observed the glow shift based on the charge of the battery, as shown in figure 3.3.2.

After all these, Thomson concluded that the glow was the negatively charged elements that came out of the first metal plate. He called these *corpuscles*, which were renamed *electrons* by physicists later. Electrons are the carriers of negative charges inside an atom. Thomson's experiment marked the discovery of the first subatomic particle.

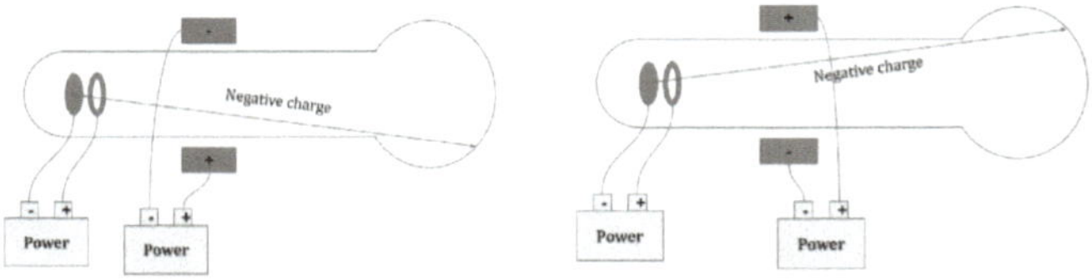

Figure 3.3.2

With the discovery of electrons inside the atom, the science of electricity got clearer. The presence of electrons also implied the existence of careers of positive charge inside the atom. Without that, the substance would not remain stable. But Thomson's experiment only showed the evidence of the electrons. Until experimental evidence revealed those positively charged particles, the atom was visualized like a *plum pudding,* as shown in Figure 3.3.3. Thomson's proposed atom model was a positively charged plum where negatively charged electrons were stuck.

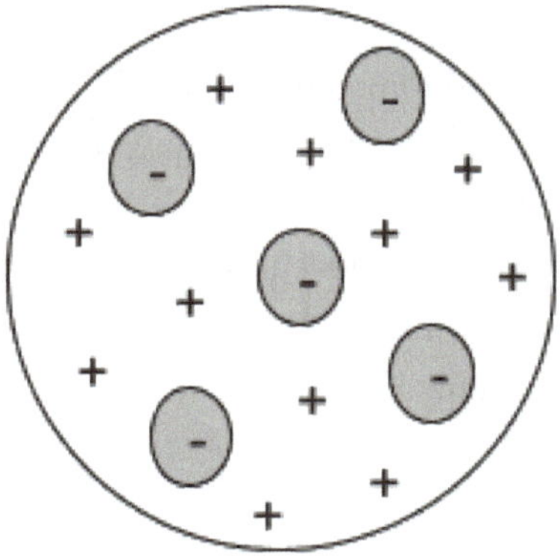

Figure 3.3.3

Almost ten years later, at the beginning of the twentieth century, 1909 to be precise, JJ Thomson's plum pudding model was challenged by Ernest Rutherford, a New Zealand born British physicist. Under the direction of Ernest Rutherford, physicists Hans Geiger and Ernest Marsden conducted the famous Gold Foil experiment that completely altered the perception of the atomic model. There were two key takeaways from that experiment.

1. The gold foil experiment revealed the presence of *protons* that were carriers of positive charge.
2. More importantly, the protons did not spread like how Thomson had envisioned. The protons stayed at the center of the atom, with the electrons orbiting around them.

Let us look at the experiment's details and see how Rutherford arrived at the above conclusions. As shown in figure 3.3.4, the experiment involved placing a thin gold foil inside a curved fluorescent screen. Rutherford intended to beam the gold foil with positively charged particles and see how they behaved after hitting the gold foil. For positively charged particles, he had chosen alpha particles. You could imagine alpha particles as those that result from the decay of radioactive elements like Radon.

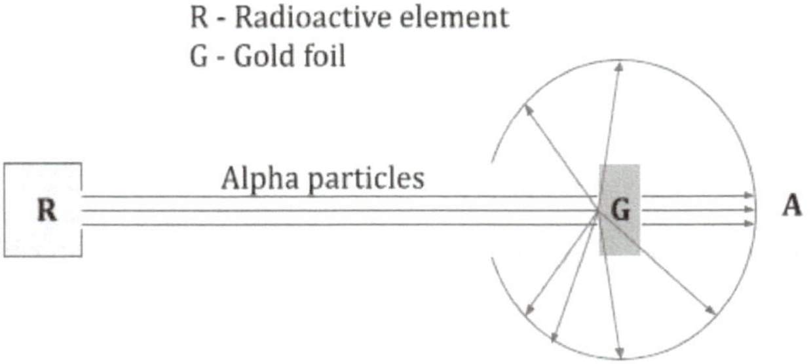

Figure 3.3.4

The first observation Rutherford had was that most of the alpha particles passed through the gold foil to hit the wall of the fluorescent screen straight (marked A in figure 3.3.4). This observation revealed that the atom was empty primarily and broke the theory of the plum pudding atom model. The second observation was even more startling. Rutherford observed random deflections of the alpha particles on the fluorescent screen after hitting the gold foil. The pass-through and the random deflections of the alpha particles allowed Rutherford to make the following inferences:

1. Inside the atom, all positive charge is concentrated at its center as a nucleus, allowing a lot of free space. The carriers of this positive charge were named *protons*. The presence of free space explained why most of the alpha particles could pass through the gold foil.
2. The electrons, on the other hand, were not static. They are more likely to be spinning around the nucleus.

From the above inferences, Rutherford proposed a nuclear structure to an atom, shown in figure 3.3.5. Rutherford's *nuclear structure* fully dethroned Thomson's plum pudding model of an atom and has remained more or less intact since then.

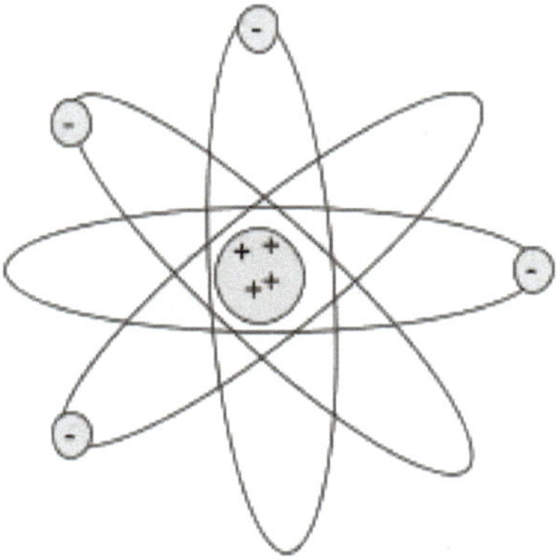

Figure 3.3.5

We say it remained more or less intact because Rutherford's nuclear structure did undergo a refinement after two decades. What led to that refinement? There is no doubt that Rutherford's proposal explained an atom's behavior better than Thomson's plum pudding model. But it was not perfect, and it had a few wrinkles.

For the atom to be stable, the number of protons would need to match the number of electrons. For multiple protons to stay as a nucleus, something has to hold them together. Else, they would repel apart. The other wrinkle was the discrepancy observed in the atom's weight, which was more than double the weight of its electrons. Physicists were able to explain this discrepancy through the concept of the neutrally charged *neutron*. Through a series of experiments, James Chadwick proved the presence of neutrons inside an atom in the year 1932. Figure 3.3.6 depicts the refined atomic structure.

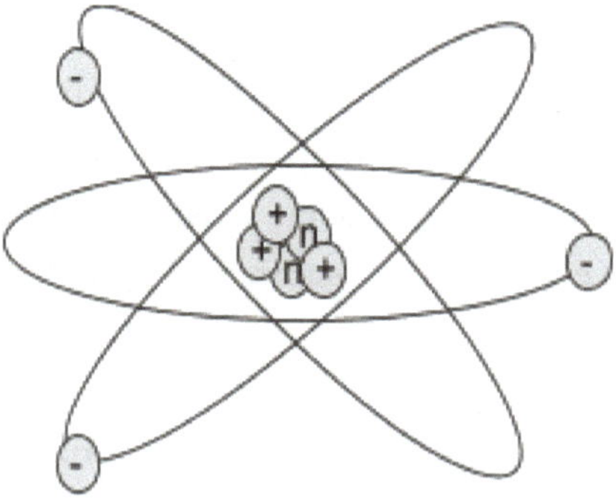

Figure 3.3.6

Let us summarize what we have learned so far. Matter comprises indivisible parts called atoms. Atoms contain *protons, neutrons,* and *electrons.* Protons are carriers of a positive charge, electrons are carriers of a negative charge, and neutrons are neutral. Per Rutherford's atomic structure, the atom's nucleus comprises protons and neutrons with negatively charged electrons spinning around the nucleus.

When an atom has the same number of protons and electrons, it remains neutrally charged. When the number of protons is higher than the number of electrons, it remains positively charged, and when the electrons are higher, it remains negatively charged. Some substances have more protons when compared to others. For instance, a neutrally charged hydrogen atom has one proton (and one electron), while a neutrally charged iron atom has 26 protons, and a neutrally charged gold atom has 79 protons. When two substances (or atoms) with varying numbers of electrons come in touch with each other, electrons tend to flow from the one that has it in excess to the one that has it less. We can thus explain electricity as the flow of electrons. For some substances (like metals), their electrons are less attached to their nucleus when compared to electrons in other substances. Electrons in such substances are freer to move around. We call such substances good conductors of electricity.

We learned that positively charged protons attract negatively charged electrons and would repel other protons. The same behavior would apply to electrons too. Given these facts, let us try a thought experiment. What would happen when we place two point charges q_1 and q_2 (it doesn't matter if it is either proton or electron) close to each other? A point charge is a hypothetical (stationary) charge located at a single point in space. The tendency to attract or repel will manifest as a force between the two charged particles, correct? That force that exists between two charged particles is called *electrostatic force* or *Coulomb Force*. The region surrounding the two particles under the influence of this force is called the electric field. The electrostatic force is named after the French scientist Charles-Augustin de Coulomb, who, in 1785, measured electrostatic force between two charged particles.

Coulomb was able to experimentally prove that the electrostatic force between two point charges is directly proportional to the product of the magnitude of the charges and inversely proportional to the square of the distances between them. This relationship is Coulomb's law, and we can mathematically state it as shown below:

$$F \propto \frac{|q_1 q_2|}{r^2}$$

$$F = |F_{21}| = |F_{12}| = k\frac{|q_1 q_2|}{r^2} \qquad (3.3.1)$$

In the above equation,

q_1 and q_2 are signed magnitudes of the point charges
r is the scalar distance between the two point charges
F is the force along the straight line joining the two point charges q_1 and q_2
k is the Coulomb's constant
F_{xy} is the force of q_x *on* q_y, as depicted in figure 3.3.7.

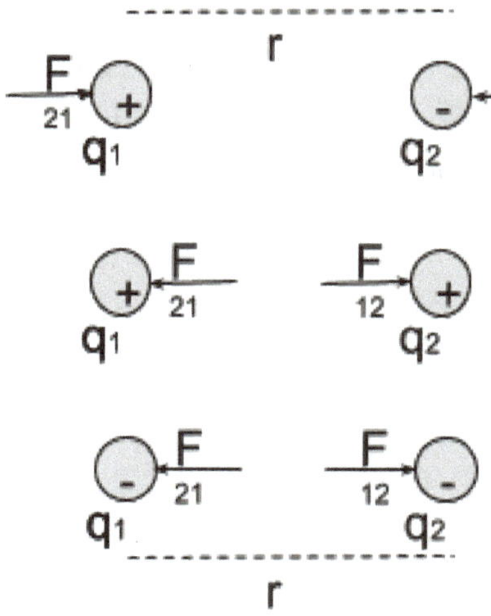

Figure 3.3.7

If the charges q_1 and q_2 have the same sign, the electrostatic force is repulsive; if they are different, the force between them is attractive. We can define Coulomb's constant k as below:

$$k = \frac{1}{4\pi\epsilon}$$

To understand the intuition behind the above expression, we will have to split the denominator of the above expression into two parts - 4π and ϵ. 4π is the angle around a point in three-dimensional space, and this division is because we are only interested in the force along the straight line that joins the two charges q_1 and q_2. Dividing by 4π allows us to rationalize the electrostatic force along a straight line. The second part ϵ is the permittivity of the medium. The idea of permittivity is that any medium (air, vacuum, etc.) is likely to resist the spread of electric fields. The constant ϵ quantifies that resistance. So, permittivity will be inversely proportional to the electrostatic force. ϵ_0 is the permittivity of free space, and we can write the Coulomb's constant in free space as:

$$k = \frac{1}{4\pi\epsilon_0}$$

$$F = |F_{21}| = |F_{12}| = \frac{1}{4\pi\epsilon_0}\frac{|q_1 q_2|}{r^2} \qquad (3.3.2)$$

You could draw parallels between Coulomb's law and Newton's law of Gravitation (equation 2.2.5). Coulomb's law is a significant milestone that also marked the starting point to the theory of electromagnetism. Also, the basic unit of charge is coulomb. One unit of positive charge is +1 coulomb which is about 6×10^{18} protons. Similarly, a negative unit of charge is -1 coulomb which is about 6×10^{18} electrons.

The flow of electricity, i.e., the rate of flow of charge carriers (protons or electrons), is called current, and the basic unit of current is ampere. When we say the current flowing through a wire is 1 ampere, what we mean is everywhere in the wire, a charge of 1 coulomb flows per second.

The last concept that we shall touch upon in electricity is voltage. For the current to start flowing, electrons need a push to start moving, which requires work. In Thomson's experiment, a battery accomplished that work. The negative charge discharged from the battery repelled the electrons in the first plate to reach the rightmost end and eventually escape from the plate inside the vacuum tube. In other words, the battery had an electric potential to discharge a certain amount of charge. Voltage is a measure of that electric potential, and we measure voltage in volts. When we say a battery has 1.5 volts (or 1.5V), what we mean is the battery has the energy to perform 1.5 joules of work to discharge 1 coulomb of charge. How would this differ from a 3V battery? A 3V battery can perform more work and discharge more coulombs of charge in the same units of time, i.e., produce more current if the medium's resistance remains the same.

So far, we have covered all the critical concepts related to electricity, and we are now ready to understand the science of light through Maxwell's equations. Maxwell's equations will be our topic of interest in the next chapter, and we shall begin that with a brief history of magnetism.

3.4 Electromagnetism

Prerequisites: To better understand this chapter's content, along with the topics shared in the earlier chapters as prerequisites, I would recommend a refresher on the following subject(s): *Calculus*. We have covered these topics in the appendix section.

A Greek legend states that magnets were first discovered accidentally by a shepherd named Magnes, who lived in Magnesia, Greece. Magnes observed the iron in his stick getting stuck to a magnetic rock. The word magnet is likely to have originated from the name of the shepherd or the place.

Both electric force and magnetic force attract other matter. Like the electric field, the surrounding region where a magnet exerts its influence is the magnetic field. Like positive and negative electric charge behavior, magnets exhibit similar behavior through the north and south poles. While the north and south poles attract, two north poles and two south poles would repel each other. Despite all these similarities, they were perceived to be unrelated until the eighteenth century.

In the year 1820, Hans Christian Oersted, a Danish scientist, as part of his experiment, discovered that the electric current that was passing through a wire from a battery began to deflect the needle of a compass that was nearby. It became clear that the two forces were manifestations of the same phenomenon, giving birth to the idea of electromagnetism.

In the nineteenth century, many physicists tried to explain the relation between electricity and magnetism or the phenomenon of electromagnetism. Key contributors in the space were Michael Faraday, André-Marie Ampère, and James Clerk Maxwell. While Ampere, Faraday, and other scientists pursued the explanation of specific aspects of electromagnetism, Maxwell took that a bit further to arrive at a unified theory that holistically explained the phenomenon of electromagnetism. Maxwell assimilated the various laws formulated by Faraday, Ampere, and others and made certain refinements to arrive at the complete field theory. It is through these equations that he proved light is an electromagnetic wave. Of the four fundamental forces, electromagnetism is the second fundamental force next to gravity. Many consider Maxwell to be the man who changed the world. When Albert Einstein was asked if he stood on the shoulders of Newton, he replied, '*No, on the shoulders of Maxwell.*' We shall now look into each of the four equations in detail:

Let us start with Coulomb's law that we just learned. If we have two point charges, q_1 and q_2, we can calculate the electrostatic force (F) between them through the following equation:

$$F = |F_{21}| = |F_{12}| = \frac{1}{4\pi\epsilon_0}\frac{|q_1 q_2|}{r^2} \qquad (3.4.1)$$

While the above equation captures the magnitude of the force, we know that force is a vector quantity, i.e., the force has a direction. So, we can accurately represent the electrostatic force between two point charges through the following equation:

$$\vec{F} = \vec{F_{21}} = \vec{F_{12}} = \frac{1}{4\pi\epsilon_0}\frac{q_1 q_2}{r^2}\hat{n}$$

In the above equation, vector $\hat{n}$ is the unit vector along the line connecting q_1 and q_2. If we consider q_2 to be 1 coulomb and represent its position as P, then the force exerted by q_1 at point P could be defined as:

$$\vec{F} = \vec{F_{12}} = \frac{1}{4\pi\epsilon_0}\frac{q_1}{r^2}\hat{n}$$

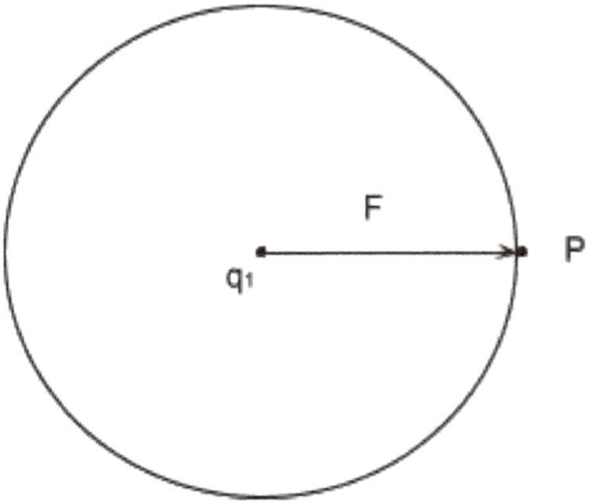

Figure 3.4.1

In figure 3.4.1, the force (attractive or repulsive) exerted by q_1 along the line towards P can also be called the *electric field intensity* of q_1 along the line towards P. We shall denote electric field intensity using the letter E.

$$\vec{E} = \frac{1}{4\pi\epsilon_0}\frac{q_1}{r^2}\hat{n} \qquad (3.4.2)$$

The force exerted by q_1 is radial, i.e., q_1's influence is in all directions. If we consider an imaginary sphere around q_1 of radius r and calculate the total force exerted on that imaginary sphere, what would be that entire force? If you recall from the *Calculus* chapter in Appendix, the idea of flux means the same thing. In the context of an electric field, we shall call it *electric flux*.

How do we calculate the total force? If we consider point P to be an infinitesimal part of that sphere *dS*, then the total electric flux would be an integration of $\vec{E}$ over the closed surface of the sphere. So, as shown below, we could equate the electric flux (ϕ_E) to the total electrostatic force (or total electric intensity) exerted on the imaginary sphere.

$$\begin{aligned}\phi_E &= \oint_S \vec{E}\cdot\hat{n}dS \\ &= \oint_S \frac{q_1}{4\pi\epsilon_0 r^2}\hat{n}\cdot\hat{n}dS\end{aligned}$$

$\hat{n}\cdot\hat{n}$ is 1 as $|\hat{n}||\hat{n}|\cos 0$ is 1.

$$\begin{aligned}\phi_E &= \frac{q_1}{4\pi\epsilon_0 r^2}\oint_S dS \\ &= \frac{q_1}{4\pi\epsilon_0 r^2}(4\pi r^2) \\ &= \frac{q_1}{\epsilon_0} \qquad (3.4.3)\end{aligned}$$

Charge q_1 is contained within the imaginary sphere. If we represent the density of charge (coulombs per cubic meter) by ρ, then the total charge q_1 can also be defined as the following volume integral:

$$q_1 = \oint_V \rho dV$$

We can rephrase equation 3.4.3 as:

$$\phi_E = \oint_S \vec{E} \cdot \hat{n} dS = \frac{1}{\epsilon_0} \oint_V \rho dV \qquad (3.4.4)$$

Using the Divergence theorem, we could equate the surface integral of the above equation to the following:

$$\oint_S \vec{E} \cdot \hat{n} dS = \oint_V \nabla \cdot \vec{E} dV = \frac{1}{\epsilon_0} \oint_V \rho dV \qquad (3.4.5)$$

$$\oint_V \nabla \cdot \vec{E} dV - \frac{1}{\epsilon_0} \oint_V \rho dV = 0$$

$$\oint_V \left(\nabla \cdot \vec{E} - \frac{\rho}{\epsilon_0} \right) dV = 0$$

$$\nabla \cdot \vec{E} - \frac{\rho}{\epsilon_0} = 0$$

$$\nabla \cdot \vec{E} = \frac{\rho}{\epsilon_0} \qquad (3.4.6)$$

The above equation captures the relation between electric charge (density ρ) and the strength and distribution (through divergence) of the electric field they create. What we have stated here is Maxwell's first equation.

Before we move further, let us bring out another concept related to electric flux from our recent derivations. The idea that we are about to learn will come in handy soon. In equation 3.4.3, we had derived total flux to the following:

$$\phi_E = \oint_S \vec{E} \cdot \hat{n} dS = \frac{q_1}{\epsilon_0}$$

Rearranging the above equation, we get:

$$\oint_S \epsilon_0 \vec{E} \cdot \hat{n} dS = q_1$$

In the above equation, the component $\epsilon_0 \vec{E}$ is called the *electric flux density*, denoted by the letter D. The surface integral of the electric flux density gives us the total charge.

$$\oint_S \vec{D} \cdot \hat{n} dS = q_1 \qquad (3.4.7)$$

$$\vec{D} = \epsilon_o \vec{E}$$

Let us take a moment to visualize the electric field created by an electric charge with the magnetic field. The first part of figure 3.4.2 shows the attractive electrostatic force between a positive point charge q_1 and a negative point charge q_2. Similarly, the second part indicates the magnetic field surrounding the north and south poles of the magnet.

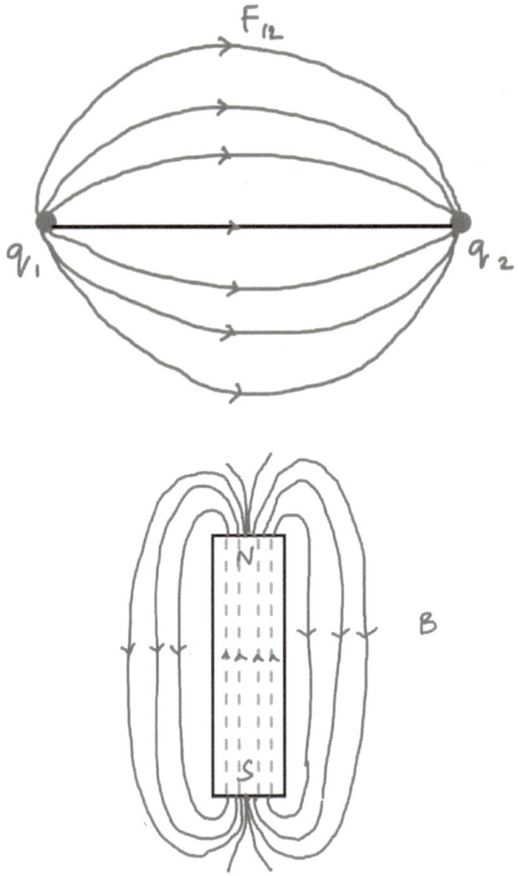

Figure 3.4.2

As you might observe, unlike the electric field where there are clear sources (or sinks) of electric charge, the magnetic field lines always flow around in closed circles across the north and the south pole of the magnet. In other words, unlike a carrier of electric charge that can either be positive or negative, a magnet's north and south pole always stay together. When broken into two, a magnet with a north and south pole will also have a north and south pole and continue as we break the magnet into even smaller pieces. However, when you break a positively charged object into two, both will remain positively charged. Let us see how we could mathematically represent this.

Let us take a closed surface in or around the magnet along its magnetic field lines and calculate its flux. Since there is no source or sink at any point, the flux would equate to zero on any closed surface. That is because the magnetic field lines that flow into that closed surface would eventually flow out. We have depicted this behavior in figure 3.4.3. Mathematically, we could calculate magnetic flux as:

$$Magnetic\,flux\,\phi = \oint_S \vec{B} \cdot \hat{n} dS = 0$$

Again, through the Divergence Theorem, we could equate the above equation to the following volume integral:

$$\oint_S \vec{B} \cdot \hat{n} dS = \oint_V \nabla \cdot \vec{B} dV = 0$$

$$\nabla \cdot \vec{B} = 0 \qquad (3.4.8)$$

What we have just arrived at is Maxwell's second equation. This equation captures the most crucial property of magnets: *There can never be magnetic monopoles.*

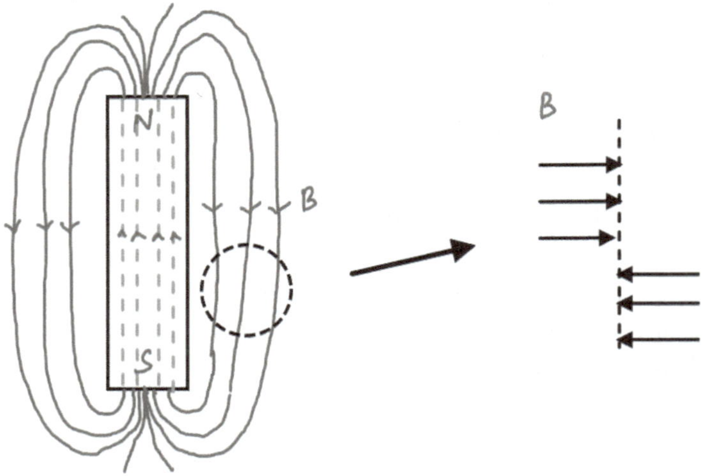

Figure 3.4.3

As you might have observed, the first two equations describe the fundamental properties of electricity and magnetism, i.e., how they both behave independently. The third and the fourth equations that we are about to see will help explain how they interact and affect.

We learned that Hans Christian Oersted first observed the relation between electricity and magnetism in 1820. However, Michael Faraday, a British scientist, captured this relation as a mathematical equation. Along with the photo of James Clerk Maxwell, Albert Einstein also had Faraday's on his study's wall.

Through a series of experiments, Faraday precisely captured the relationship between magnetic field and electricity. As shown in figure 3.4.4, the essence of the experiment he conducted is as follows: He had a coil of wire (why the coil will be made clear) connected to a galvanometer. A galvanometer is an instrument used to detect and measure the flow of current. Understandably the galvanometer read zero current in the beginning. But after passing a bar magnet through the coil, he found the current flow to happen for a moment that showed up as a blip in the galvanometer. He also observed the intensity of the current increasing with an increase in the number of coils. Also, when he performed the experiment by inserting the opposite pole of the same magnet, he observed the same amount of current flowing but in the opposite direction.

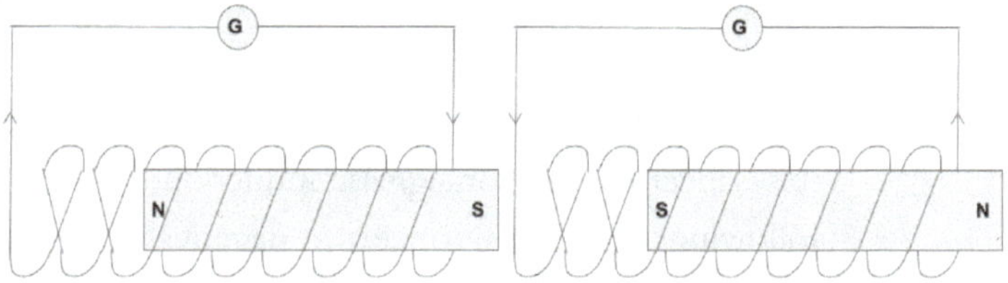

Figure 3.4.4

Faraday arrived at the following conclusions:

1. A change in the magnetic field created an electric potential (voltage) that resulted in the flow of current through the coil. That induced electric potential is called the *electromotive force* or *emf*, which we represent by the letter *e*. This entire phenomenon is called *electromagnetic induction.*
2. The current was produced in the coil only when the magnet moved relative to the coil. In other words, to produce current, the intensity of the magnetic field

has to change over time. Magnetic flux (ϕ_B) can help quantify the intensity of the magnetic field, as shown in figure 3.4.5.

$$\text{Rate of change of intensity in magnetic field} = \frac{\partial \phi_B}{\partial t}$$

3. The polarity and the direction of the magnet's movement influenced the direction of the current flow through the coil. In the year 1834, Emil Lenz, a Russian scientist, got precise in capturing this relationship. Lenz deduced that the induced emf produced current that opposes the change of magnetic flux that produced it. Lenz's law captures this aspect of electromagnetic induction through the following equation. Notice the negative sign added to the previous equation.

$$e = -\frac{\partial \phi_B}{\partial t}$$

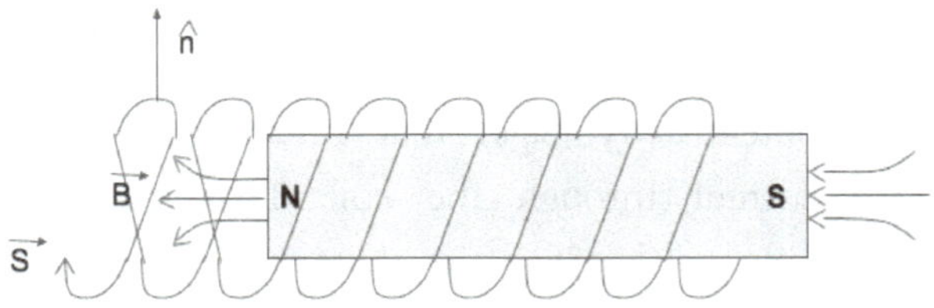

Figure 3.4.5

The electromotive force induced in the coil results from some electric field $\vec{E}$ created by the change in magnetic field $\vec{B}$. If we represent the coil as a curved path $\vec{S}$, we could define the electromotive force along the coil through the following line integral.

$$e = \int_C \vec{E} \cdot d\vec{S}$$

Using Stoke's theorem, we could write the above equation as the following surface integral:

$$e = \int_C \vec{E} \cdot d\vec{S} = \int_S \nabla \times \vec{E} \cdot \hat{n} dA$$

In the equation above, *dA* is an infinitesimal part of the coil's surface when considered a cylinder. As shown in figure 3.4.5, $\hat{n}$ is the unit vector perpendicular to the surface of the coil.

In similar lines, we could rewrite the right-hand side of Faraday's equation to the following surface integral:

$$-\frac{\partial \phi_B}{\partial t} = -\frac{\partial}{\partial t} \int_S \vec{B} \cdot \hat{n} dA$$

With the above, we could restate Faraday's law as the following:

$$\int_S \nabla \times \vec{E} \cdot \hat{n} dA = -\frac{\partial}{\partial t} \int_S \vec{B} \cdot \hat{n} dA$$

$$\int_S \nabla \times \vec{E} \cdot \hat{n} dA + \frac{\partial}{\partial t} \int_S \vec{B} \cdot \hat{n} dA = 0$$

$$\int_S \left(\nabla \times \vec{E} \cdot \hat{n} + \frac{\partial}{\partial t} \vec{B} \cdot \hat{n} \right) dA = 0$$

$$\nabla \times \vec{E} \cdot \hat{n} + \frac{\partial}{\partial t} \vec{B} \cdot \hat{n} = 0$$

$$\nabla \times \vec{E} \cdot \hat{n} = -\frac{\partial}{\partial t} \vec{B} \cdot \hat{n}$$

$$\nabla \times \vec{E} = -\frac{\partial}{\partial t} \vec{B} \qquad (3.4.9)$$

With the above, we have arrived at Maxwell's third equation. Now, let us move on to Maxwell's last equation.

While Faraday was exploring the impact of magnetic fields on electricity, around the same time, Ampere was investigating the effects that electric current can have on magnetic fields. Building on the findings of Orsted, Ampere discovered that a magnetic field encircled a current-carrying wire. He proposed measuring the total strength of that magnetic field that encircled the wire in Amperes/meter. In other words, the total magnetic strength (B_{Total}) surrounding a meter-long wire carrying I Amperes could be equated to I.

$$B_{Total} = I$$

The above equation holds good only in principle because we are yet to factor in the effect of the medium. Similar to the idea of permittivity (ϵ) that we had in the electric field, we shall introduce the notion of *permeability* (μ) in the context of the magnetic field. Though both permittivity and permeability measure the influence of the medium, there is a crucial difference. Permittivity is how much the medium resists the flow of the electric field, and permeability is how much the medium allows the flow of the magnetic field. So, permittivity is inversely proportional to electric flux, and permeability is directly proportional to magnetic flux. After considering the influence of the medium, we could restate the previous equation as:

$$B_{Total} = \mu I$$
$$B_{Total} = \mu_0 I \; (\mu_0 \; is \; the \; permeability \; of \; free \; space) \qquad (3.4.10)$$

As shown in figure 3.4.6, if $\vec{B}$ represents the magnetic field at a particular point surrounding the wire, we could define the total magnetic field as the following line integral. Vector $\vec{S}$ represents the circular path.

$$B_{Total} = \oint_C \vec{B} \cdot d\vec{S}$$

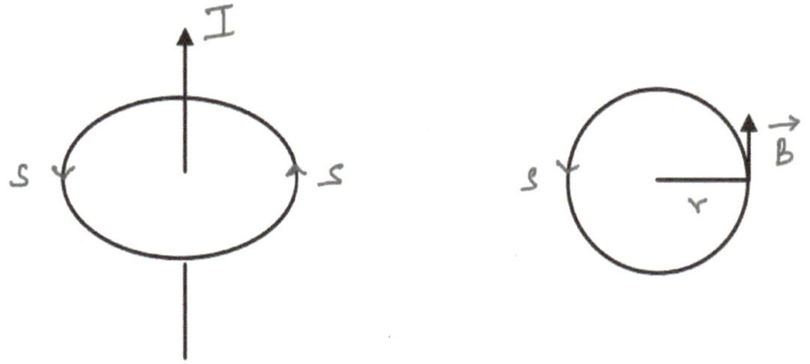

Figure 3.4.6

Using Stoke's theorem, we could write the above equation as the following surface integral:

$$B_{Total} = \oint_C \vec{B} \cdot d\vec{S} = \oint_S \nabla \times \vec{B} \cdot \hat{n} dA \qquad (3.4.11)$$

In equation 3.4.11, *dA* is an infinitesimal part in the magnetic lines' surface when considered an imaginary cylinder. Also, $\hat{n}$ is the unit vector perpendicular to the surface of the imaginary cylinder. On the other hand, the current density could also express the total current (*I*). As shown in figure 3.4.7, if you cut across the wire, you could visualize the current density to vary from center to edge across the concentric circles. In other words, we could envision the spread of current density as a flux of current. The idea of current density is similar to how we calculated electric flux density in equation 3.4.7. If we represent the current density by vector $\vec{J}$, we could calculate the total current using the following surface integral.

$$I = \oint_S \vec{J} \cdot \hat{n} dA$$

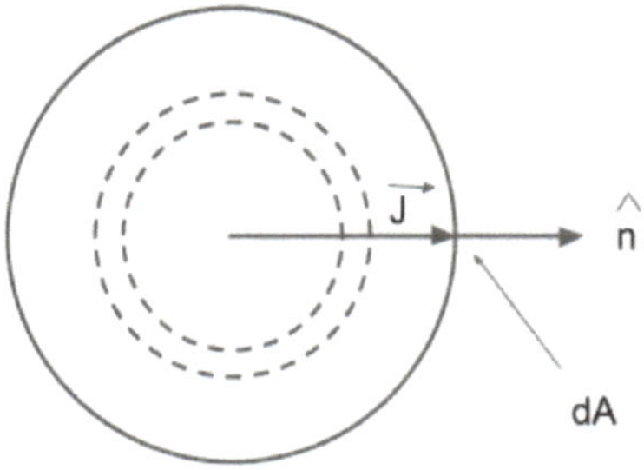

Figure 3.4.7

Replacing the above expression for I in equation 3.4.10 and through the equivalence of equations 3.4.10 and 3.4.11, we arrive at the following Ampere's law:

$$\oint_S \nabla \times \vec{B} \cdot \hat{n} dA = \mu_0 \oint_S \vec{J} \cdot \hat{n} d\vec{A}$$
$$\nabla \times \vec{B} = \mu_0 \vec{J}$$

The above equation is Ampere's circuital law. This equation, as proposed by Ampere, was accurate but had a limitation. The equation meant current flow produced a magnetic field. However, experiments also revealed other ways to create magnetic fields. We have shown one such setup in figure 3.4.8. In the circuit, we have connected a specific voltage to a capacitor. The capacitor has two metal plates and a clear gap in between. A gap meant there was no way current could flow across the two plates. However, we will observe the magnetic field created around the gap too. A magnetic needle would deflect the same way at points A and B. It was clear to Maxwell that it was not only current that created a magnetic field. So, Maxwell proposed a refinement to Ampere's equation.

$$\nabla \times \vec{B} = \mu_0 \left(\vec{J} + \vec{J'} \right)$$

Along with regular current $\vec{J}$, Ampere proposed another component $\vec{J'}$ that is capable of creating magnetic fields. Ampere called that component *Displacement current.* Taking cues from Faraday's law, Maxwell proposed the rate of change of the electric flux or, to be precise, the rate of change of electric flux density to be the cause of the magnetic field between the two plates of the capacitor.

$$\vec{J'} = \frac{\partial \vec{D}}{\partial t}$$

We know $\vec{D} = \epsilon_0 \vec{E}$. So,

$$\vec{J'} = \epsilon_0 \frac{\partial \vec{E}}{\partial t}$$

Replacing $\vec{J}'$ in Maxwell's refined equation, we get:

$$\nabla \times \vec{B} = \mu_0 \vec{J} + \mu_0 \epsilon_0 \frac{\partial \vec{E}}{\partial t} \qquad (3.4.12)$$

What we have arrived at is Maxwell's fourth equation. In free space, there will not be any current but only displacement current. In that scenario, J would be 0, and Maxwell's fourth equation reduces to the following:

$$\nabla \times \vec{B} = \mu_0 \epsilon_0 \frac{\partial \vec{E}}{\partial t} \qquad (3.4.13)$$

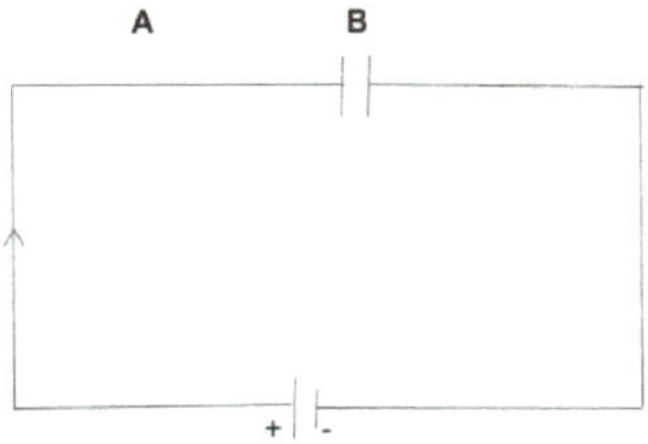

Figure 3.4.8

To summarize, the four Maxwell equations are:

$$\nabla \cdot \vec{E} = \frac{\rho}{\epsilon_0}$$
$$\nabla \cdot \vec{B} = 0$$
$$\nabla \times \vec{E} = -\frac{\partial}{\partial t}\vec{B}$$
$$\nabla \times \vec{B} = \mu_0 \vec{J} + \mu_0 \epsilon_0 \frac{\partial \vec{E}}{\partial t}$$

In free space, i.e., in an environment free of any charge or current, we could restate Maxwell's equations as shown below:

$$\nabla \cdot \vec{E} = 0$$
$$\nabla \cdot \vec{B} = 0$$
$$\nabla \times \vec{E} = -\frac{\partial}{\partial t}\vec{B}$$
$$\nabla \times \vec{B} = \mu_0 \epsilon_0 \frac{\partial \vec{E}}{\partial t}$$

It is through the above four equations that Maxwell unified the theory of electricity and magnetism. It is for this contribution that we hail Maxwell as the man who changed the world. Maxwell's genius does not stop here. His real genius lies in what he did with these four equations. In the next chapter, we shall see how Maxwell explained the science of light through these equations.

3.5 Deriving the speed of light

Prerequisites: To better understand this chapter's content, along with the topics shared in the earlier chapters as prerequisites, I would recommend a refresher on the following subject(s): *Symmetries in Vector Calculus* & *Wave Equation.* We have covered these topics in the appendix section.

I have a confession to make. The excitement to learn everything about Einstein's mass-energy equation led me to the path of learning. However, in that journey, my most significant discovery was James Clerk Maxwell. It is while writing this chapter that my admiration and respect for Maxwell spiked. What set Maxwell apart from his peer scientists was that he was a remarkable mathematician. During the nineteenth century, many physicists pursued the search for truth through experiments. However, Maxwell tried to make sense of the world through mathematics.

More than once, Maxwell showed something through math that the world was yet to see. In 1856, when Maxwell was around 25 years old, he published an essay on the stability of Saturn rings that won him the Adams prize. Galileo first observed Saturn's rings at the beginning of the seventeenth century. The exact science behind the rings that appeared like a solid structure through the telescope had no concrete explanation back then. In his essay, Maxwell hypothesized the ring comprised unconnected particles, with each particle acting as a satellite of Saturn. Many of those independent moving particles created an illusion to the human eye that it is a solid structure. He had backed his theory with mathematical calculations. Space expeditions that happened much later proved Maxwell correct.

Maxwell did something similar with the four equations. His brilliance did not end with him arriving at the unified theory but how he chose to connect them all. If you observe the four equations, the first two explained the nature of electricity and magnetism, and the last two equations explained how they affected one another. The third equation stated that the rate of change of magnetic field would manifest as an electric field that would curl in the neighborhood. The fourth equation said something very similar, but from the standpoint of the electric field, i.e., the rate of change of electric charge density would manifest as a magnetic field that would curl in the neighborhood.

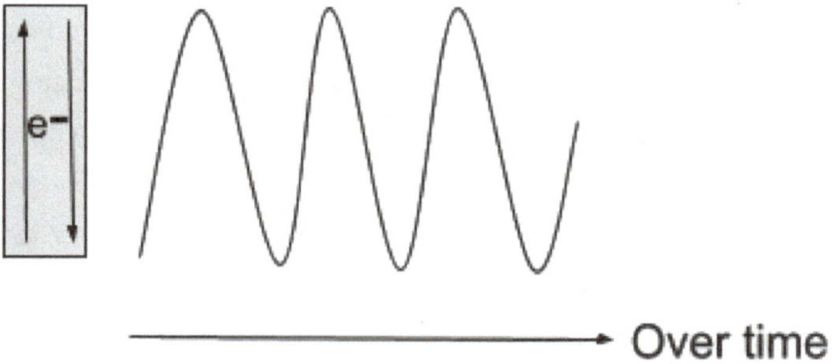

Figure 3.5.1

With these four equations, Maxwell performed an interesting thought experiment. He thought, how would a continuously oscillating electron affect its neighborhood? In other words, if an electron accelerates from zero to a certain speed, then decelerates back to zero and does that in a loop (as shown in figure 3.5.1), how would it affect its surroundings? The second part of figure 3.5.1 shows the varying intensity of the electric field plotted against time. Maxwell's fourth equation tells us that such a rate of change in electric field intensity would manifest as a magnetic field. Owing to that constant acceleration and the deceleration of the electron, the resulting magnetic field will also vary. That would manifest as an electric field as per Maxwell's third equation. This recursive behavior or one (magnetic field) affecting the other (electric field) should technically continue as long as the electron oscillates.

Maxwell hypothesized that an oscillating electron, through the above-mentioned recursive behavior, should propagate something into its neighborhood. He wasn't sure what that was. Had such a thought occurred in the mind of Faraday, he would have proceeded with an experiment. Being a mathematician, Maxwell chose to solve it through math. He thought: *If the rate of change of magnetic field manifests as an electric curl* ($\nabla \times \vec{E}$)*, which in turn manifests as a magnetic field, let me see what the curl of* ($\nabla \times \vec{E}$) *is.* He pursued to calculate the following:

$$\nabla \times \left(\nabla \times \vec{E}\right)$$

Let us derive the above equation, the way Maxwell proceeded back then. We can expand the above expression in two different ways. This first expansion is from Maxwell's third equation (equation 3.4.9), and the second expansion is the symmetry we derived in vector calculus. Remember, everything around this oscillating electron is free space. So, we would consider the variation of Maxwell's equations in free space. We have provided both the expansions below.

$$\nabla \times \left(\nabla \times \vec{E}\right) = -\nabla \times \frac{\partial \vec{B}}{\partial t} = -\frac{\partial}{\partial t}\left(\nabla \times \vec{B}\right)$$
$$\nabla \times \left(\nabla \times \vec{E}\right) = \nabla\left(\nabla \cdot \vec{E}\right) - \nabla^2 \vec{E} = -\nabla^2 \vec{E}$$

Equating the two expanded expressions, we get

$$\nabla^2 \vec{E} = \frac{\partial}{\partial t}\left(\nabla \times \vec{B}\right)$$
$$\nabla^2 \vec{E} = \frac{\partial}{\partial t}\left(\mu_0 \epsilon_0 \frac{\partial \vec{E}}{\partial t}\right)$$
$$\frac{\partial^2 \vec{E}}{\partial t^2} = \frac{1}{\mu_0 \epsilon_0} \nabla^2 \vec{E}$$
$$\frac{\partial^2 \vec{E}}{\partial t^2} = \left(\frac{1}{\sqrt{\mu_0 \epsilon_0}}\right)^2 \nabla^2 \vec{E} \qquad (3.5.1)$$

As you might recall, the above is a wave equation. Maxwell concluded that an oscillating electron would propagate a wave and called them *electromagnetic waves*. We can visualize an electromagnetic wave, as shown in figure 3.5.2. Notice how the two fields, electric (E) and magnetic (B), propagate perpendicular (as per the fourth equation) to one another. As per the above equation, the speed of the electromagnetic wave equates to the following:

$$\frac{1}{\sqrt{\mu_0 \epsilon_0}}$$

After replacing μ_0 and ϵ_0 with the actual values, the above calculation yielded a value that precisely matched the speed of light. That astonishing coincidence acted as a huge clue that made Maxwell conclude that light has to be an electromagnetic wave.

Like the Saturn rings, Maxwell predicted that electromagnetic waves could be created in a laboratory setup but didn't demonstrate them experimentally. Almost two decades after Maxwell predicted the existence of electromagnetic waves, in 1887, Heinrich Hertz, a brilliant German physicist and experimentalist, successfully created radio waves in his homemade laboratory. Radio waves are a form of electromagnetic wave. The experiment confirmed that electromagnetic waves, as predicted by James Clerk Maxwell, do exist.

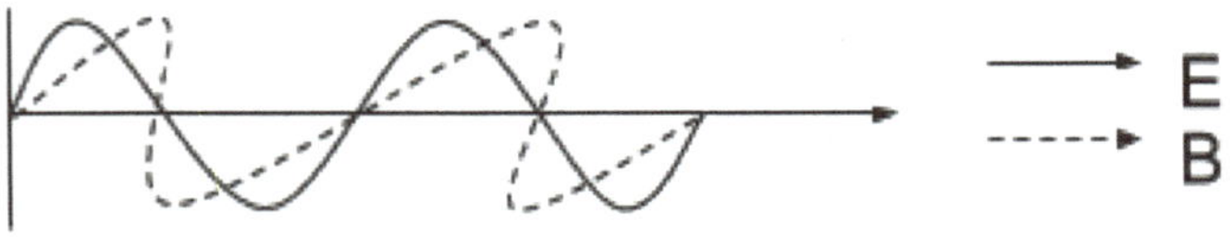

Figure 3.5.2

4. TIMELESS LIGHT

4.1 The conundrum

We have reached the point of understanding and stating the problem about time fully. The entire section 2 was about the science of light, and our motivation to understand the science of light was born from the last thought experiment we had in section 1. The thought experiment involved the lightning scenario as perceived by two individuals: one traveling inside a high-speed container and the other outside the container. Two lightning flashes that strike the two ends of the high-speed container at the exact moment would be perceived as simultaneous events (read events at the same time) by the person on the pavement but not so by the person inside a high-speed moving container. Einstein called this *Relativity in Simultaneity* in his writing. That discrepancy arose due to the moving container. Inside the moving container, light ended up having to travel a shorter distance from one end of the container post the lightning strike. In other words, there is no *absolute time* that both people (A and B) could align on. Like space, the concept of time is relative too. The conundrum is, without absolute time and absolute space, we have no way to transform an event between two frames of reference.

At the end of section 1, we briefly debated a possibility that could solve or even avoid this conundrum, i.e., could the speed of light change with the speed of the medium? Is that true? Let us try to answer that question both mathematically and experimentally. Let us start with the Math angle. The closest we have reached to understanding the science of light is through Maxwell's formula for the speed of an electromagnetic wave:

$$c = \frac{1}{\sqrt{\mu_0 \epsilon_0}}$$

If light is an electromagnetic wave, it depends on the permeability (μ) of the medium and the permittivity (ϵ) of the medium. The speed at which the medium moves is clearly out of the above equation. What about the experimental evidence? In 1851, French physicist Hippolyte Fizeau performed a vital experiment to measure the effects of a medium's movement on the speed of light. It is referred to as the Fizeau experiment and has been quoted multiple times by Albert Einstein in his writings. Through the experiment, Fizeau calculated the dragging effect that a moving medium could have on the speed of light. He found clear evidence that the speed of the moving medium had negligible influence on the speed of light. The influence was insignificant when the chosen medium was water, and when the medium was air, the impact was nil. Both math and experiments favor the possibility that the speed of light is independent of the movement of the medium. We shall refer to this behavior of light as the *Law of propagation of light.* Until we prove otherwise, we will have to conclude that both time and space are relative.

To summarize our learnings so far, the lack of absolute space led to the idea of multiple frames of reference. We then used Galilean transformation to transform from one frame of reference to another. Remember, the Galilean transformation would require time to be absolute. We have now realized that the idea of time is not absolute either. How are we then going to transform an event from one frame of reference to another? It is now a perfect moment to indulge in the genius of Einstein.

4.2 The idea of spacetime

Through the law of the propagation of light, we have realized that we neither have absolute space nor absolute time. Though events occur at the same moment (like lightning), they do not happen at the same moment in time because of the distance light has to travel in the two frames of reference. The frames of reference here are the moving container and the stationary pavement. As we discussed in one of the previous chapters, time is only a counter on a mechanical device (aka *clock*) that we have created. To pinpoint an event in time, we would have to perceive that event, and for us to perceive the event, light has to travel a distance to reach our eyes.

Without absolute time, we will be unable to transform an event from one frame of reference to another through the Galilean transformation. Now, what is our way out? Einstein proposed considering space and time as one single unit and not as separate to resolve this problem. He called it *spacetime* which essentially means the idea of time is local to a frame of reference. In technical terms, in a frame of reference, the coordinates of an event will have four components - $<x,y,z,t>$ instead of $<x,y,z>$. While the idea of spacetime accommodates the reality that there is no absolute time, it does leave on our hands the task of transforming spacetime events across frames of reference. We will learn a new method called the Lorentz transformation that will help us accomplish that. To understand the Lorentz transformation, we need to clarify spacetime's implications on the distance between two points.

Let us come back to the original thought experiment that motivated us to learn the science of light. To recap, we have the situation of a high-speed moving container (frame of reference is *K*) relative to a stationary pavement (frame of reference *K'*). Person A is inside the container and person B is on the pavement. Let us consider two points inside the container *P* and *Q*. Our pursuit is to calculate the distance between the two points from the perspective of A (frame of reference *K*) and B (frame of reference *K'*).

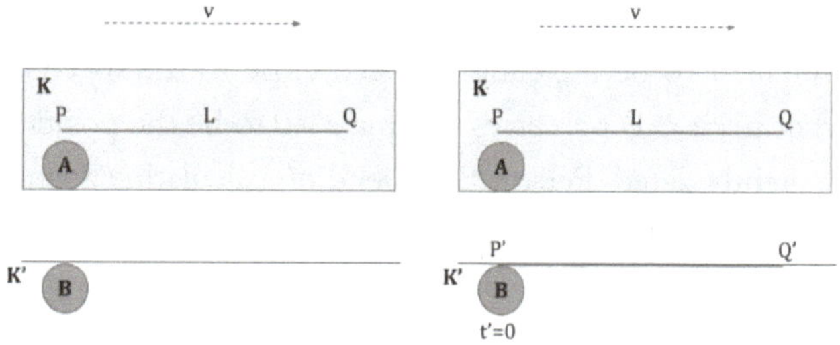

Figure 4.2.1

Let us start with the frame of reference *K*. For person A, who is inside the container, all they would need is to take a tape and measure the distance between the points P and Q. Let us say Person A derives that value to be *L*. Now, what is the best option for person B to calculate the distance between P and Q? Remember, P and Q are points in the frame of reference *K* moving at high speed. Person B decides to take a snapshot, i.e., plot the positions of P and Q at a particular time *(t'=0)* in *K'*. The second part of figure 4.2.1

shows that the two points are marked as P' and Q' in the pavement. Now, person B could use a tape to measure the same. Can person B accurately plot the two points in *K'* to arrive at the distance *L* as calculated by person A? The law of propagation of light would not allow us to get the same value. Let us understand the reason behind the same. Think of person B to be standing on the pavement waiting for the first point to cross him. The moment the first point (P) inside the container crosses him, he takes a snapshot (time *t'=0*), i.e., person B visually plots the positions of both P of Q in his frame of reference (*K'*) as *P'* and *Q'*. While person B might think that he had plotted *Q'* as shown in figure 4.2.1, in reality, he would have plotted *Q'* as shown in figure 4.2.2. Since Q is relatively far compared to *P* and given the container to be traveling at speed close to the speed of light, what person B perceives at time *t'=0* to be the position of Q is slightly dated. Person B will end up calculating a length (*L'*) that will be shorter than L.

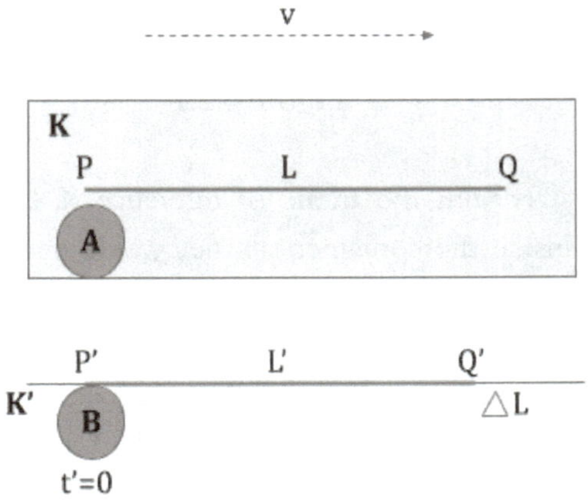

Figure 4.2.2

This effect is called *length contraction* or *Lorentz contraction* (or sometimes Lorentz-FitzGerald contraction). It is the phenomenon of a moving object's length appearing shorter than its length measured in the object's rest frame. Again, the idea of movement and rest in the above statement is relative. To be precise, the distance measured between two points on the pavement will appear shorter when plotted inside the container. It is because from the perspective of the container, the container is at rest (inertia is undisturbed in the container), and it is the pavement that is moving with a velocity of v. We learned this in section 1 from Newton's first law of motion, i.e., law of inertia. Since the relative velocity is the same, i.e., v in this case, the contracted length from the perspective of K is the same as the contracted length from the perspective of K'.

With the details we were able to cover, we are ready to derive the formula for Lorentz transformation that we shall use to transform events in spacetime across frames of reference.

4.3 Lorentz transformation

Like Galilean transformation, the idea of Lorentz transformation will help transform events between two frames of reference: *K* and *K'*. *K'* is moving close to the speed of light along a straight line relative to *K*. Spacetime coordinates *(x,y,z,t)* represent an event in *K*, and coordinates *(x', y', z', t')* represent an event in *K'*. There is no absolute space but absolute time in Galilean transformation, and in Lorentz transformation, there is neither absolute space nor absolute time. In other words, $t \neq t'$ in the Lorentz transformation.

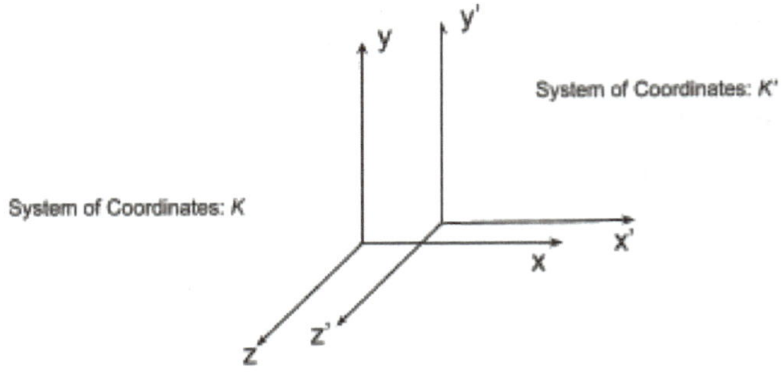

Figure 4.3.1

Since *K'* is moving along a straight line, we can assume the *x-axes* of both the frames of reference to coincide. We can see this in figure 4.3.1. Given these, a transformation of an event from *K* to *K'* would only involve calculating *x'* and *t'* for a given *x* and *t*. In other words, for an event *(x, y, z, t)* in

K, its transformation in *K'* will have the same values for *y'* and *z'*.

$$y' = y$$
$$z' = z$$

Let us consider a light signal that travels at a speed of *c* in the direction of the x-axis. In the frame of reference *K*, we could represent the distance traveled by light through the following equation.

$$x = ct$$
$$x - ct = 0$$

We can interpret the above equation as all events perceived at the spacetime coordinate (x, t) through the frame of reference *K*. If the above equation is non-zero, then all those events would either be in the past or the future in *K*.

An equivalent of the earlier equation in the frame of reference *K'* is provided below. Remember, the directions of *x* and *x'* coincide:

$$x' = ct'$$
$$x' - ct' = 0$$

Let us keep the equations together before we proceed.

$$x - ct = 0$$
$$x' - ct' = 0$$

Given that one of the frames of reference moves at a constant velocity *v* along a straight line and the speed of light is a constant, events as perceived as present in one frame should shift by a fixed spacetime in the other frame. For the above statement to hold good, there should be a constant γ such that,

$$\gamma = \frac{x - ct}{x' - ct'} \ or \ x - ct = \gamma(x' - ct')$$

γ will be 1 when $v=0$. The above equations hold good for the light signal traveling along the positive direction of the x-axis. We could represent scenarios where the light signal travels along the negative direction using another constant.

$$\mu = \frac{x + ct}{x' + ct'} \ or \ x + ct = \mu(x' + ct')$$

Adding (and subtracting) the above two equations, we could arrive at the following equations to calculate x' and ct'

$$x' = ax - bct \qquad (4.3.1)$$
$$ct' = act - bx \qquad (4.3.2)$$

As you might notice, we have introduced two additional constants to keep the equations simple.

$$a = \frac{\gamma + \mu}{2}$$
$$b = \frac{\gamma - \mu}{2}$$

In the frame of reference *K'*, the origin is *x'=0*. We could equate that origin to the following coordinate in *K*. We arrive at the following by replacing *x' = 0* in equations 4.3.1.

$$x = \frac{bc}{a}t$$

$$\frac{x}{t} = \frac{bc}{a}$$

The above derivation tells us that the origin of the frame of reference *K'* moves at a velocity of *bc/a* in the frame of reference *K*. In other words, we could say *K'* moves with a velocity *v*, that equates to the following:

$$v = \frac{bc}{a}$$

Now, let us try and calculate the contraction in length as observed from each of the two frames. We can calculate the contraction in the length of an object in *K'* from *K* by taking its snapshot in *K*. We could do this by equating *t* to *0* in equation 4.3.1.

$$x' = ax$$

$$x = \frac{x'}{a}$$

When *x'=1*,

$$\Delta x = \frac{1}{a}$$

We have derived that an object of length 1 unit in *K'* will contract by *1/a* units when measured in *K*. Similar to the above, we can also calculate the contraction in the length of an object in *K* from *K'*. We can do this by equating *t'* to *0* in equation 4.3.2. Snapshot from *K'* i.e., when *t'=0.*

$$t = \frac{bx}{ac}$$

Substituting the above in equation 4.3.1,

$$x' = ax - \frac{b^2 x}{a}$$
$$x' = a\left(1 - \frac{b^2}{a^2}\right)x$$
$$x' = a\left(1 - \frac{v^2}{c^2}\right)x$$

When *x=1,*

$$\Delta x' = a\left(1 - \frac{v^2}{c^2}\right)$$

We have derived that an object of length 1 unit in *K* will contract by $a(1-(v^2/c^2))$ unit when measured in *K'*. Given that the inertia is undisturbed within *K* and *K'* and the relative velocity is a constant between *K* and *K'*, the contraction in length as observed from the two frames of reference would be the same. We learned this towards the end of the last chapter. In other words,

$$\Delta x = \Delta x'$$

Equating the values for $\Delta x \; and \; \Delta x'$, we get:

$$a^2 = \frac{1}{\left(1 - \frac{v^2}{c^2}\right)}$$

$$a = \frac{1}{\sqrt{1 - \frac{v^2}{c^2}}}$$

Solving for *b*, we get:

$$b = \frac{v}{c}\left(\frac{1}{\sqrt{1 - \frac{v^2}{c^2}}}\right)$$

Substituting the above values for *a* and *b* in equations 4.3.1 and 4.3.2, we get:

$$x' = \frac{x - vt}{\sqrt{1 - \frac{v^2}{c^2}}}$$

$$t' = \frac{t - \frac{v}{c^2}x}{\sqrt{1 - \frac{v^2}{c^2}}}$$

To summarize, for a spacetime coordinate *(x,y,z,t)* in *K*, its Lorentz transformation in *(x', y', z', t')* in *K'* is:

$$x' = \frac{x - vt}{\sqrt{1 - \frac{v^2}{c^2}}} \qquad (4.3.3)$$

$$y' = y \qquad (4.3.3)$$

$$z' = z \qquad (4.3.3)$$

$$t' = \frac{t - \frac{v}{c^2}x}{\sqrt{1 - \frac{v^2}{c^2}}} \qquad (4.3.3)$$

4.4 Special Theory of Relativity

Let us summarize our learning. Galileo pioneered the idea of relativity (Galilean relativity) in 1632 when it became apparent that there would not be a single frame of reference. In other words, there is no absolute space. Galileo stated that all the laws of motion are the same in all inertial frames. An inertial frame is either a moving frame or a frame at rest, whose inertia is undisturbed. In the later years, what Galileo referred to as the laws of motion was formalized by Sir Isaac Newton (that we now call Newton's laws of motion or Newtonian mechanics).

As discussed earlier, Galileo articulated the idea of relativity through the example of a ship traveling at a constant velocity on a smooth sea. He stated, as long as the ocean remains smooth and the ship does not rock, any observer below the deck would not be able to tell whether the ship was moving or stationary. That specific frame of reference (aka system of coordinates) that the laws of motion hold relative to is called the *Galilean system of coordinates*. Thus, was born the technique of Galilean transformation that helped transform an event from one Galilean system of coordinates to another. In other words, through the method of Galilean transformation, we transform an event from one frame of reference to another.

Galilean transformation had a vital presupposition. The presupposition is the presence of absolute time. With

Maxwell's discovery of light being an electromagnetic wave, whose velocity remained constant irrespective of the movement of the medium, an apparent incompatibility emerged between Galilean relativity and electromagnetism. Galilean transformation worked when one frame of reference traveled at a smaller velocity relative to the other. Through the lighting example in section one, we saw it fall short when the frame of reference moved at speed close to the speed of light.

Albert Einstein corrected the above to arrive at a more accurate theory called the special theory of relativity. Albert Einstein proposed all the concepts discussed in this section, be it the idea of spacetime and the Lorentz transformation of spacetime coordinates across frames of references, in the year 1905 as the *Special Theory of Relativity*. This theory was the foundation for the other paper, titled *Does the Inertia of a Body Depend Upon Its Energy Content?* which Albert Einstein published in the same year. It is in the second paper that Einstein derived the mass-energy equation. The core aspects of Einstein's special theory of relativity are the concept of spacetime and Lorentz transformation. More formally, Einstein based his theory on the following postulates:

1. The laws of physics are invariant, i.e., identical in all inertial frames of reference.
2. The speed of light in a vacuum is the same for all observers. In other words, light travels at a constant speed irrespective of the motion of the light source or the observer.

Einstein further generalized his special theory of relativity and published a much-refined version of the same in 1915. The refined version is what we now call the *General Theory of relativity*. Understanding the general theory of relativity is not required to derive the mass-energy equation, so we might as well save that for another book.

5. THE CLIMAX

5.1 Deriving e=mc²

Prerequisites: To better understand this chapter's content, along with the topics shared in the earlier chapters as prerequisites, I would recommend a refresher on the following subject(s): *Binomial Theorem*. We have covered these topics in the appendix section.

We have reached the climax. This chapter will put all the knowledge we have gained to derive Einstein's mass-energy equation. To begin with, let us consider two frames of reference, *K* and *K'*. The frame of reference *K'* moves at a constant velocity *v* relative to *K*. We represent a spacetime event in *K* by <*x,y,z,t*> and a spacetime event in *K'* by <*x', y', z', t'*>. Similar to the considerations that we held in the Lorentz transformation; we can assume the *x-axes* of both the frames of reference coincide.

Given the above, let us consider a light wave traveling at speed *c* along the *x-axis* in frame *K*. If *L* is the energy acquired by that light wave after traveling a unit distance in *K*, what would be its equivalence in the frame of reference *K'*? We had learned that energy and work are two sides of the same coin, and work is force times the distance traveled (or displacement). Force is mass times the acceleration as per Newton's second law. Light being an electromagnetic wave, travels with the same velocity *c* in both *K* and *K'*. So, there is no additional acceleration that it experiences in *K'* relative to

K. Also, since the frame *K'* is free of any acceleration, the mass of the light ray would measure the same in both *K* and *K'*. If you are not convinced, you could try this experiment to weigh yourself on the floor and a moving elevator. Except for the moment when the elevator begins to move, i.e., when acceleration is non-zero, your weight would remain the same at all moments. What about the distance traveled (or displacement) by the light ray in *K*? We learned that one unit in *K* would not be the same in *K'*. Applying the Lorentz transformation, if we denote the equivalence of *L* in *K'* as *L'*, then we could calculate *L'* through the following formula.

$$L' = L\frac{1-\frac{v}{c}}{\sqrt{1-\frac{v^2}{c^2}}} \qquad (5.1.1)$$

In the above equation, we have used the Lorentz transformation (equation 4.3.3) to derive *x'* that corresponds to *x=1* in *K*. We shall be using the above equation soon.

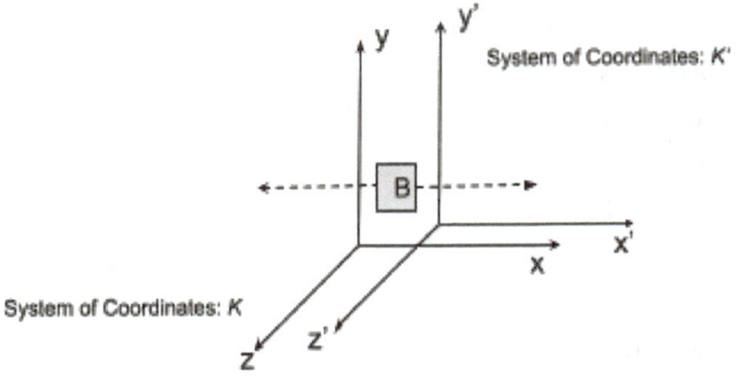

Figure 5.1.1

Let there be a stationary object (B) in K that has an energy E_0. Let H_0 be the energy of the body relative to K'. Let us consider this body to send out plane waves of light, i.e., emit equal quantities of light simultaneously in the opposite direction, as shown in figure 5.1.1.

The law of conservation of energy should hold good in K and K'. In other words, the body's energy after emitting the light wave (E_1 and H_1) should be the difference between the body's energy before emitting the light wave and the energy acquired by the light wave. The energy acquired by the light wave is intentionally split into two parts since it travels in opposite directions.

$$
\begin{aligned}
E_0 &= E_1 + \frac{L}{2} + \frac{L}{2} \\
H_0 &= H_1 + \frac{L'}{2} + \frac{L'}{2}
\end{aligned}
$$

We can use equation 5.1.1 to replace L' and expand the above equation. Just to remind, the below equations consider the energy acquired by a light wave that traveled a unit distance in the frame of reference K.

$$
\begin{aligned}
E_0 &= E_1 + \frac{L}{2} + \frac{L}{2} \\
H_0 &= H_1 + \frac{L}{2}\left(\frac{1+\frac{v}{c}}{\sqrt{1-\frac{v^2}{c^2}}}\right) + \frac{L}{2}\left(\frac{1-\frac{v}{c}}{\sqrt{1-\frac{v^2}{c^2}}}\right) \\
&= H_1 + \frac{L}{\sqrt{1-\frac{v^2}{c^2}}}
\end{aligned}
$$

Notice the change in sign as we replace *L'*. The reason is that the frame of reference *K'* is moving in one direction, but the light-emitting body emits equal quantities of light simultaneously in opposite directions. Subtracting E_0 from H_0, we get the following.

$$H_0 - E_0 - (H_1 - E_1) = L\left(\frac{1}{\sqrt{1-\frac{v^2}{c^2}}} - 1\right) \qquad (5.1.2)$$

H and *E* are energy values of the same body referred to two systems of coordinates, *K* and *K'*, which are in motion relative to each other. Since the body is in motion, the difference H − E includes the kinetic energy *K* plus the other forms of energy. So far, we don't have any evidence of the other forms of energy being any different across two frames of reference whose inertia is undisturbed. For instance, the gravitational potential energy of a hanging iron ball will be the same in both *K* and *K'*. If gravity is in the direction of the *y* axis, we know through the Lorentz transformation that *y'=y*. So, we shall consider the other forms of energy to be a constant C across *K* and *K'*. In other words,

$$H_0 - E_0 = K_0 + C$$
$$H_1 - E_1 = K_1 + C$$

We could write equation 5.1.2 as:

$$K_0 - K_1 = L\left(\frac{1}{\sqrt{1-\frac{v^2}{c^2}}} - 1\right)$$

If the difference in the kinetic energy is non-zero, then there has to be a difference to the body's mass given the other variable *v*, i.e., velocity is a constant. If *m'* denotes the difference in the body's mass after emitting the light wave, we could restate the above equation as:

$$\frac{1}{2}(m')v^2 = L\left(\frac{1}{\sqrt{1-\frac{v^2}{c^2}}} - 1\right)$$

$$\frac{1}{2}(m')v^2 = L\left(\left(1-\frac{v^2}{c^2}\right)^{-\frac{1}{2}} - 1\right)$$

Replacing the right-hand side of the above equation with the generalized binomial expansion, we get the following.

$$\frac{1}{2}(m')v^2 = L\left(\left(1+\frac{1}{2}\frac{v^2}{c^2}+\ldots\right) - 1\right)$$

We could approximate the binomial expansion to the first two components as the remaining components approach zero. Given that, we could equate the above equation to the following:

$$\frac{1}{2}(m')v^2 = \frac{1}{2}L\frac{v^2}{c^2}$$

$$m'c^2 = L$$

Here it is, and we have derived Einstein's mass-energy equation. We have provided a more popular form of this equation below.

$$e = mc^2$$

If a body gives off the energy *e* in the form of radiation, its mass diminishes by e/c^2.

6. APPENDIX

6.1 Vectors

Let us consider this scenario: We have kept an empty backpack that weighs 500 grams (.5 kgs) on a square table. We need to place two objects, say a laptop (weighing 1 kg) and a book (weighing .5 kgs) in the backpack. We will assume that the backpack is big enough to accommodate the two.

If we were to answer the question: *What will be the net weight of the backpack after placing the laptop and the book?* Our answer would involve a simple calculation as below.

The new weight of the backpack = weight of the backpack + weight of laptop + weight of the book

New weight of the backpack = .5+ 1+.5

New weight of the backpack = 2 kgs

Now, let us take a slightly different scenario. Let us say we apply some force and push the backpack at two points simultaneously. The first push is given from the bottom (A in figure 6.1.1) and the other from the side (B). The SI unit for force is *Newton.* So let us quantify the force applied from the bottom to be *5 Newtons* and the force applied on the side to be *3 Newtons.* Now, if we ask a similar question: *What would be the net force on the backpack?* What would our response be? For the moment, let us exclude the effects of gravity.

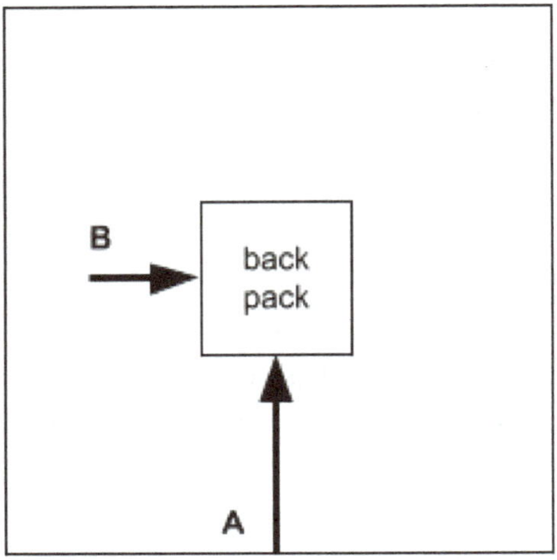

Backpack on a table

Figure 6.1.1

The net force acting on the backpack is not a simple addition of forces *A and B* until we apply the two forces in the same direction. It is clear from the above figure that we have not applied the two forces in the same direction. The notation of 5 Newtons (for force *A*) is incomplete as it captures only the magnitude of the force. It does not give a clue about the force's direction unless accompanied by the picture in Figure 6.1.1. Before we begin to calculate the resultant net force, we need a notation that allows us to capture both the magnitude and the direction of the two forces.

What is the notation? Now visualize the table as a cartesian coordinate system (as shown in figure 6.1.2) whose center coincides with the backpack's center. Let us notate each unit on the *x-axis* using the variable *i* and each unit on the *y axis* using variable *j*. In this example, each unit on both the axes corresponds to one Newton.

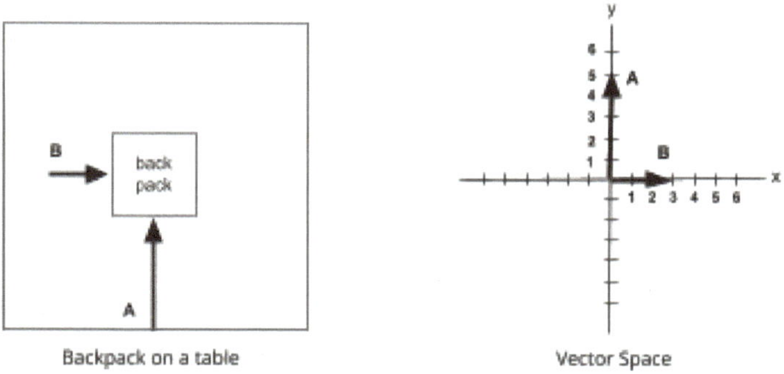

Figure 6.1.2

What we have just defined is called the *Vector space.* In this vector space, we can represent forces *A and B* as vectors:

$$\vec{A} = 0\vec{\imath} + 5\vec{\jmath}$$
$$\vec{B} = 3\vec{\imath} + 0\vec{\jmath}$$

Why did we include the arrow signs ($\rightarrow$) on top of the variables? It is only to denote that these are special variables in the vector space. What is the advantage of the above notation? This notation allows us to capture both the magnitude and the direction of the two forces. Also, in a short while, we will see how this notation will enable us to perform a few other interesting and useful operations. We calculate the magnitude of any vector $\vec{V}$ (represented as $a\vec{\imath} + b\vec{\jmath}$) using the formula $\sqrt{a^2 + b^2}$. We denote the magnitude of a vector $\vec{V}$ as $|\vec{V}|$. You might notice that the calculation is similar to calculating the distance between two points in Euclidean geometry. Using this formula, we could calculate the magnitude of the two forces A and B as follows:

$$|\vec{A}| = \sqrt{0^2 + 5^2} = 5\,Newtons$$
$$|\vec{B}| = \sqrt{0^2 + 3^2} = 3\,Newtons$$

As you have observed, a vector representation captures both the direction and the magnitude of the force. Once we plot the forces on a vector space, we could add the two vectors to calculate the net force. For the two vectors A and B, we can calculate its net force $\vec{C}$ as shown below:

$$\vec{C} = \vec{A} + \vec{B} = 0\vec{\imath} + 5\vec{\jmath} + 3\vec{\imath} + 0\vec{\jmath} = (0+3)\vec{\imath} + (5+0)\vec{\jmath} = 3\vec{\imath} + 5\vec{\jmath}$$
$$|\vec{C}| = \sqrt{3^2 + 5^2} = \sqrt{34} = 5.83 \sim 6\,Newtons$$

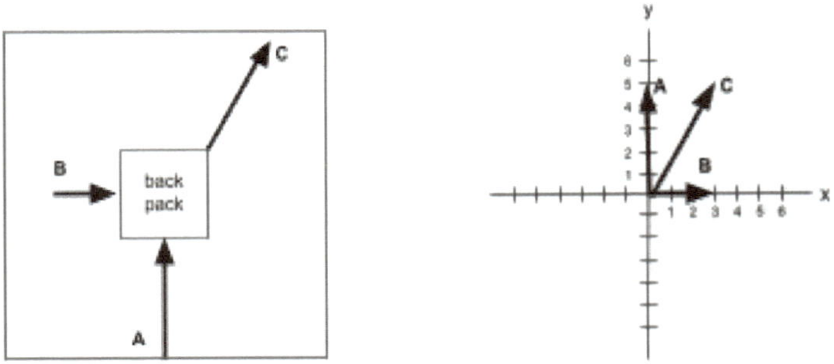

Backpack on a table

Vector Space

Figure 6.1.3

The net force acting on the backpack $\vec{C}$ is depicted in figure 6.1.3, both visually and in the vector space. We could visualize the additive nature of the three vectors in Figure 6.1.4.

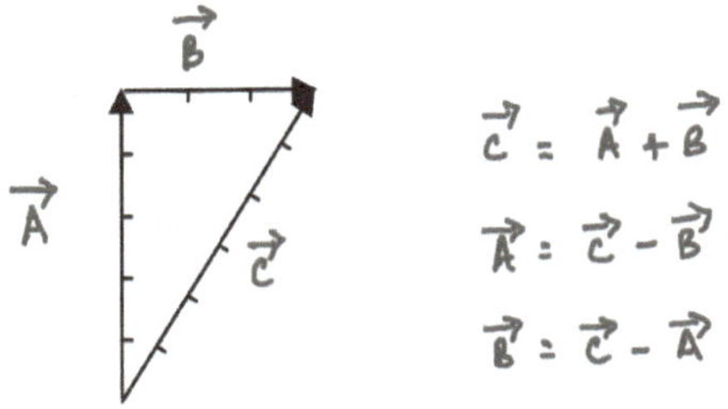

Figure 6.1.4

While the examples seen above are in a 2-dimension vector space, we can easily extend this to a 3-dimension vector space by adding the unit vector $\vec{k}$ for the *z-axis*. While we may not visualize beyond three dimensions, there is no limit mathematically, and we can have any number of dimensions in a vector space. We have shown a vector ($\vec{D}$) in the 3-D vector space in Figure 6.1.5.

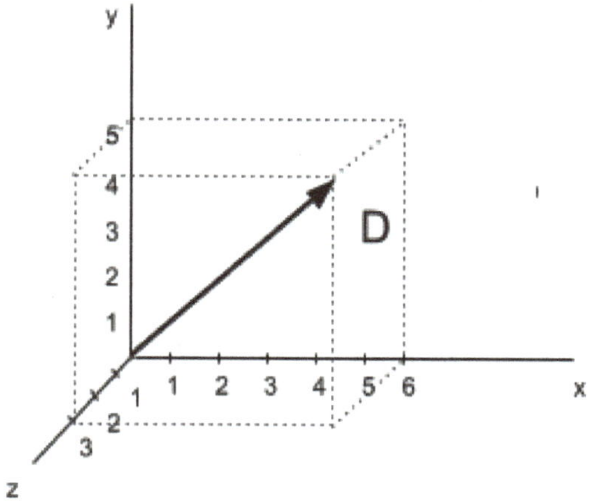

Figure 6.1.5

$$\vec{D} = 6\vec{\imath} + 5\vec{\jmath} + 3\vec{k}$$

Not just force, vector space can help represent any phenomenon with the following two aspects: magnitude and direction. Examples include the movement of water, electric charge, the velocity of an object, etc.

Let us now look at the other interesting manipulations that we could do with vectors, and for that, we will bring back to our original backpack example. We will do a small thought experiment, and in that experiment, we shall neither increase nor decrease the magnitude of the two forces *A and* B. Now, let us think about how we should direct both forces to push the backpack off the table in the quickest time? It might look intuitive to you that we can maximize the magnitude of the resultant force *C* when we apply the two forces *A and B* in the same direction (shown in figure 6.1.6), either from the side or from the bottom. When done so, the magnitude of the resultant force will be 8 Newtons, and the path traveled by the backpack towards the edge would be the shortest. Remember, the table is square-shaped.

$$\textit{When both forces are applied from the side}:$$
$$\vec{C} = \vec{A} + \vec{B} = 5\vec{\imath} + 0\vec{\jmath} + 3\vec{\imath} + 0\vec{\jmath} = (5+3)\vec{\imath} + (0+0)\vec{\jmath} = 8\vec{\imath} + 0\vec{\jmath}$$

$$\textit{When both forces are applied from the bottom}:$$
$$\vec{C} = \vec{A} + \vec{B} = 0\vec{\imath} + 5\vec{\jmath} + 0\vec{\imath} + 3\vec{\jmath} = (0+0)\vec{\imath} + (5+3)\vec{\jmath} = 0\vec{\imath} + 8\vec{\jmath}$$

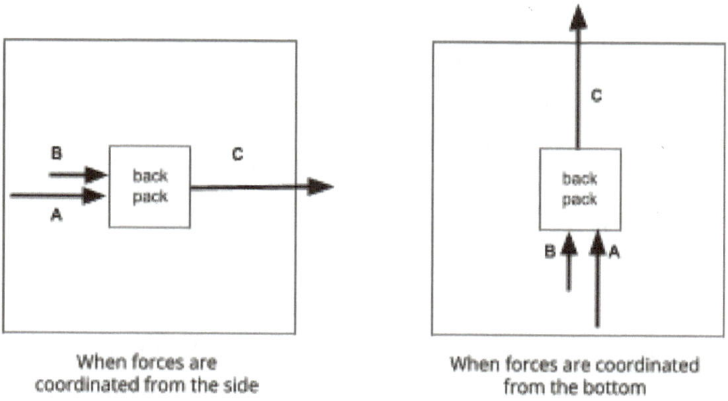

Figure 6.1.6

In many scenarios in science, it will be helpful to determine if two vectors supplement one another, i.e., if one enhances the strength of the other. For two vectors to strengthen one another, both the vectors would need to be aligned, i.e., they need to have the same direction, which would require the angle between them to be 0 degrees. The notion of the *Vector Dot product* is to determine the alignment between two vectors. We can calculate the dot product of two vectors using the formula below:

$$\vec{A} \cdot \vec{B} = \left|\vec{A}\right|\left|\vec{B}\right| \cos\theta \qquad (6.1.1)$$

The dot product of two vectors turns out to be an effective tool in physics to determine if two vectors (force, electric charge, etc.) head in the same direction. When the two vectors are in the same direction, i.e., when the angle between the vectors is 0°, the dot product would be positive ($\cos 0 = 1$). Similarly, when the two vectors are perpendicular to each other, i.e., when the angle between the vectors is $90°$, the dot product would be zero ($\cos 90 = 0$). Finally, when two vectors are in the opposite direction, the dot product would be a negative value ($\cos 180 = -1$).

The dot product of two vectors is the magnitude of the first vector times the magnitude of the projection of the second vector on the first vector. As shown in figure 6.1.7, if θ is the angle between $\vec{A}$ and $\vec{B}$, $|\vec{A}|\cos\theta$ is the projection of $\vec{A}$ on $\vec{B}$. When the vector $\vec{A}$ is parallel to $\vec{B}$, its projection on $\vec{B}$ becomes $|\vec{A}|$

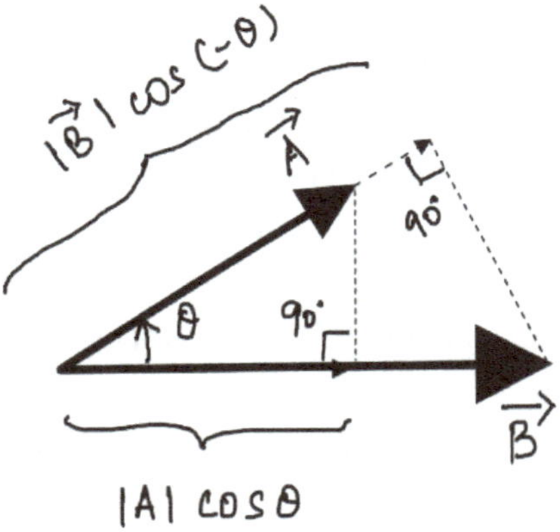

Figure 6.1.7

If $\vec{A}\cdot\vec{B}$ is $|\vec{A}||\vec{B}|\cos\theta$, then what about $\vec{B}\cdot\vec{A}$? Let us derive that as well but before we do that, let us remember that we have considered θ to be the angle between $\vec{A}$ and $\vec{B}$. This way, the angle between $\vec{B}$ and $\vec{A}$ would be $-\theta$.

$$\vec{B}\cdot\vec{A} = |\vec{B}||\vec{A}|\cos(-\theta)$$

$$\vec{B} \cdot \vec{A} = \left|\vec{B}\right|\left|\vec{A}\right| \cos\theta$$

$$\vec{B} \cdot \vec{A} = \left|\vec{A}\right|\left|\vec{B}\right| \cos\theta$$

$$\vec{B} \cdot \vec{A} = \vec{A} \cdot \vec{B}$$

We can see from the above relation that the dot product of two vectors is a commutative operation, i.e., the operands' order does not matter. You could refer to Figure 6.1.8 if you are curious to know how cos $(-\theta)$ equals $\cos\theta$.

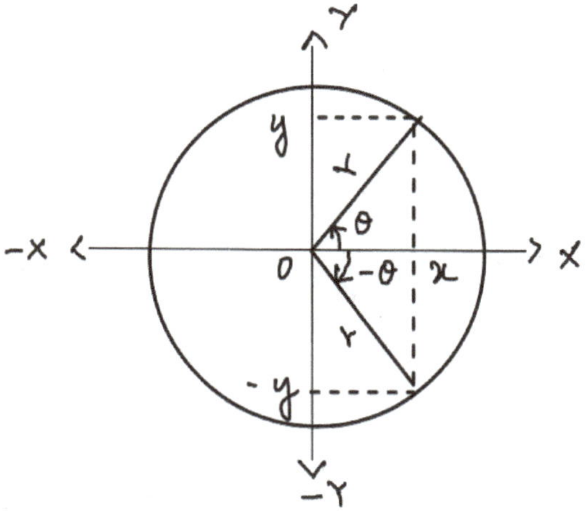

$$\cos\theta = \frac{x}{r}$$

$$\cos(-\theta) = \frac{x}{r} = \cos\theta$$

$$\sin\theta = \frac{y}{r}$$

$$\sin(-\theta) = \frac{-y}{r} = -\sin\theta$$

Figure 6.1.8

So far, we have derived the dot product of two vectors geometrically. We shall extend the above to get to its algebraic equivalent. To do that, we shall consider two things:

1. For a brief moment, we will generalize the two vectors $\vec{A}$ and $\vec{B}$ as the following:
$$\vec{A} = a_1\vec{i} + a_2\vec{j}$$
$$\vec{B} = b_1\vec{i} + b_2\vec{j}$$
2. As shown in figure 6.1.8A, we shall equate angle θ to $\alpha - \beta$

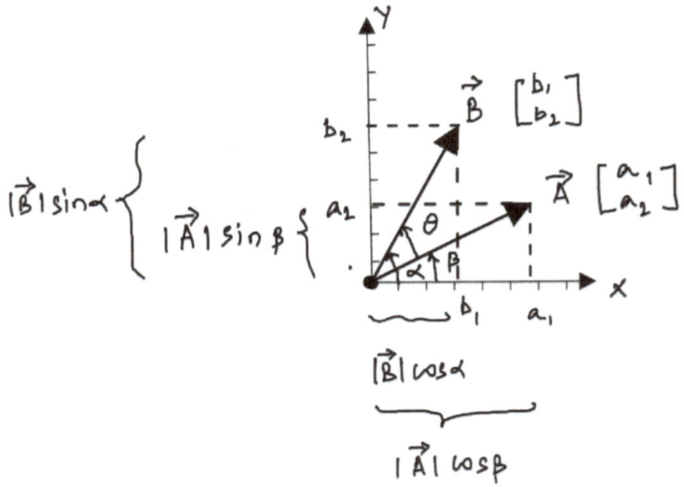

Figure 6.1.8a

$$\vec{A}.\vec{B} = \left|\vec{A}\right|\left|\vec{B}\right|\cos\theta$$
$$= \left|\vec{A}\right|\left|\vec{B}\right|\cos(\alpha - \beta)$$

From figure 6.1.8A, we can derive $\cos(\alpha - \beta) = \cos\alpha\cos\beta + \sin\alpha\sin\beta$

$$
\begin{aligned}
\vec{A}.\vec{B} &= \left|\vec{A}\right|\left|\vec{B}\right|(\cos\alpha\cos\beta + \sin\alpha\sin\beta) \\
&= \left|\vec{A}\right|\left|\vec{B}\right|\cos\alpha\cos\beta + \left|\vec{A}\right|\left|\vec{B}\right|\sin\alpha\sin\beta \\
&= \left|\vec{A}\right|\cos\alpha\left|\vec{B}\right|\cos\beta + \left|\vec{A}\right|\sin\alpha\left|\vec{B}\right|\sin\beta \\
&= \left|\vec{A}\right|\cos\beta\left|\vec{B}\right|\cos\alpha + \left|\vec{A}\right|\sin\beta\left|\vec{B}\right|\sin\alpha \\
&= a_1b_1 + a_2b_2
\end{aligned}
$$

Extending the above, we can calculate the dot product of two vectors in 3-dimensional space as follows:

$$
\vec{A}\cdot\vec{B} = \begin{bmatrix} a_1 \\ a_2 \\ a_3 \end{bmatrix} \cdot \begin{bmatrix} b_1 \\ b_2 \\ b_3 \end{bmatrix} = a_1b_1 + a_2b_2 + a_3b_3
$$

If you observe, in the formula for dot product, the cosine of the angle ($\cos\theta$) alone could get us the sense of alignment of two vectors, i.e., how parallel are the two vectors. If our need is less about magnitude but more about precisely measuring the alignment of the two vectors, could we get away with $|\vec{A}|$ and $|\vec{B}|$? Yes, we could do that, but before that, we would need to get a slice of the two vectors $\vec{A}\, and\, \vec{B}$.

So, what do we mean by a slice of a vector? It is a vector with a magnitude of 1 in the direction of the original vector. Such vectors are called *unit vectors,* and we denote them using a hat (^) on top. For vectors $\vec{A}\, and\, \vec{B}$, $\hat{u}\ and\ \hat{v}$ are its unit vectors.

$$
\begin{aligned}
\vec{A} &= 5\vec{\imath} + 0\vec{\jmath} \\
\vec{B} &= 0\vec{\imath} + 3\vec{\jmath}
\end{aligned}
$$

$$\hat{u} = \frac{\vec{A}}{|A|} = \left(\frac{5\vec{i} + 0\vec{j}}{5}\right) = (5/5)\vec{i} + (0/5)\vec{j} = \vec{i}$$

$$\hat{v} = \frac{\vec{B}}{|B|} = \left(\frac{0\vec{i} + 3\vec{j}}{3}\right) = (0/3)\vec{i} + (3/3)\vec{j} = \vec{j}$$

Once we have the unit vectors for $\vec{A}\, and\, \vec{B}$ in the form of $\hat{u}\ and\ \hat{v}$, we could only perform a dot product of $\hat{u}\ and\ \hat{v}$ to determine how parallel are the directions of vectors $\vec{A}\, and\, \vec{B}$.

$$\hat{u} \cdot \hat{v} = |\hat{u}||\hat{v}| \cos\theta$$
$$\hat{u} \cdot \hat{v} = \cos\theta \quad since\, |\hat{u}| = |\hat{v}| = 1$$

In our examples, both vectors $\vec{A}\, and\, \vec{B}$ were parallel to the *x-axis* or the *y axis*. Let us consider a new vector $\vec{X}$ that is neither parallel to the *x-axis* nor the *y-axis* and calculate its unit vector. Let us denote the unit vector of $\vec{X}$ as $\hat{w}$. We have provided the calculation of $\hat{w}$ for a specific vector $\vec{X}$ below. We have also visually depicted the two vectors in figure 6.1.9.

$$\vec{X} = 3\vec{i} + 4\vec{j}$$
$$|\vec{X}| = \sqrt{3^2 + 4^2} = 5$$
$$\hat{w} = \frac{\vec{X}}{|X|} = \left(\frac{3\vec{i} + 4\vec{j}}{5}\right) = (3/5)\vec{i} + (4/5)\vec{j} = 0.6\vec{i} + 0.8\vec{j}$$

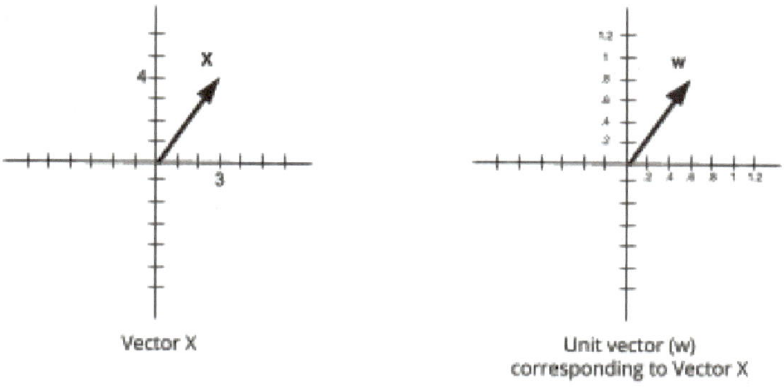

Figure 6.1.9

Apart from the dot product, there is another exciting vector operation that we shall explore now. To understand that, let us do another thought experiment. As shown in figure 6.1.10, consider two vectors A and B fully aligned with one another, i.e., their dot product is 1. Unlike the earlier example, where vectors A and B represented the force applied on a backpack, the two vectors represent the force of water flow. You could imagine A and B to be two streams of water, each flowing through an independent hose with a different force. Vector A is thicker, so let us consider A's force greater than that of B.

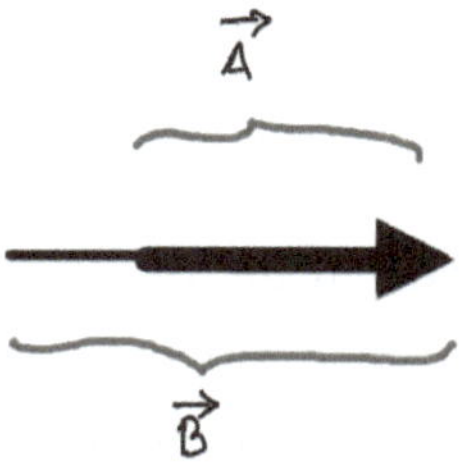

Figure 6.1.10

Let us imagine we place these hoses on the floor of your garden in the same direction. What can we expect to see when the water comes out of the two hoses? To keep things simple, we shall consider the terrain, i.e., floor (since we are dealing with water) flat and exclude the effects of gravity. Since we have placed the two hoses parallel, we could expect the water from both the hoses to continue to travel in the same direction. We can see this in Figure 6.1.11.

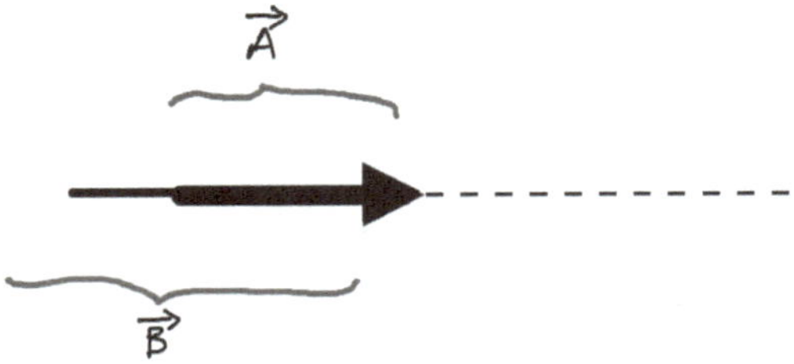

Figure 6.1.11

What would happen if we change the direction of one of the hoses, say A. Since A is a more potent force, as shown in figure 6.1.12a, it is likely to push the particles from B and make the surrounding wet, correct? What could be our estimate of the area that will be wet? There are two influencing factors here:

- The first factor is the magnitude of the forces of the flow. The greater the force, the higher the scatter.
- The second factor is the angle between the two vectors. The greater the angle, the greater the scatter. When the angle is 0, the scatter is nil, and the scatter is the highest when the angle is 90. If θ is the angle between the vector, $\mathbf{sin}\,\theta$ ($\mathbf{sin\,0}^{\circ}$ is 0 and $\mathbf{sin\,90}^{\circ}$ is 1) could be a valuable measure of this scatter.

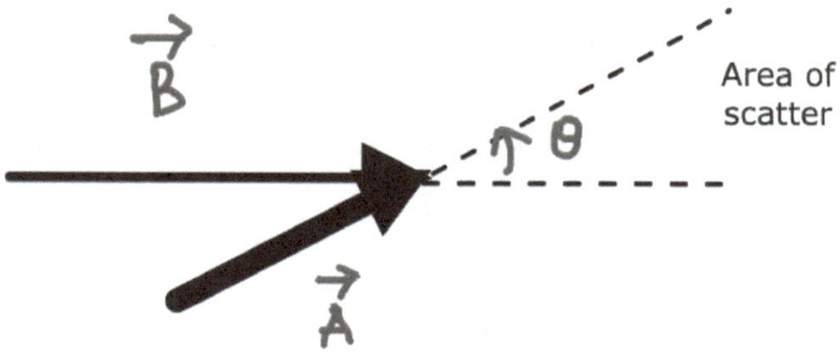

Figure 6.1.12a

So, to increase the wet area, we would need to increase the scatter, i.e., we would need to allow the water particles to obstruct one another. The obstruction happens when two vectors meet at an angle. We are ready to learn the notion of the *Vector cross product.* The cross product of two vectors $\vec{A}\, and\, \vec{B}$ is denoted as $\vec{A} \times \vec{B}$ and helps us estimate the rejection of one vector ($\vec{B}$) in the hands of the other vector. ($\vec{A}$). We *incompletely* define Cross Product through the following formula:

$$\vec{A} \times \vec{B} = \left|\vec{A}\right|\left|\vec{B}\right| \sin\theta \qquad (6.1.2)$$

Why is the above formula incomplete? To understand the incompleteness, let us consider a slightly different scenario of hose A pushing the water from the top (as shown in figure 6.1.12b). As you might imagine, the area of the scatter is likely to differ now. In the original hose positions, water spread in the counterclockwise direction. In the revised positions, water would spread in the clockwise direction. This behavior will reflect in the sign of the value derived through equation 6.1.2. Suppose we have considered the angle in the first position to be θ (the standard notation is to consider counterclockwise movement as positive). In that case, the angle in the revised position will be $-\theta$, and we know that:

$$\sin\theta = -\sin(-\theta)$$

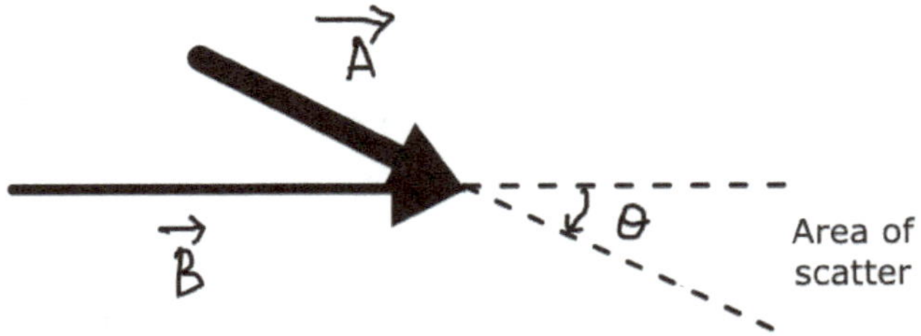

Figure 6.1.12b

If the value of the above expression captures the magnitude and the sign captures the direction of the spread, what is still not captured in the above expression is the orientation or the axis of rotation. This gap reveals a crucial aspect of a cross product, i.e., unlike dot products, the cross product of two vectors is also a vector.

So, how do we convert the formula (equation 6.1.2) for cross product into a vector, and what should that vector signify? The standard notation is to apply the right-hand rule and define the cross product in the direction of a vector $\vec{N}$ that is perpendicular to both $\vec{A}$ and $\vec{B}$. Vector $\vec{N}$ gives us the axis of rotation. Per the right-hand rule, we need to hold the right hand's thumb, index, and middle finger perpendicular to one another. If the index and the middle fingers represent

$\vec{A}\ and\ \vec{B}$ and the angle between them is counterclockwise (θ), then the direction of the thumb represents $\vec{N}$. In other words, $\vec{N}$ is orthogonal to the plane formed by $\vec{A}\ and\ \vec{B}$. If $\hat{n}$ is the unit vector in the direction of $\vec{N}$, then the complete formula for $\vec{A} \times \vec{B}$ is defined as follows:

$$\vec{A} \times \vec{B} = \left|\vec{A}\right|\left|\vec{B}\right| \sin\theta\, \hat{n}$$

The unit vector $\hat{n}$ is called the *normal vector* as it is normal (read orthogonal) to the plane formed by $\vec{A}\ and\ \vec{B}$. Figure 6.1.13 shows an instance of a normal vector.

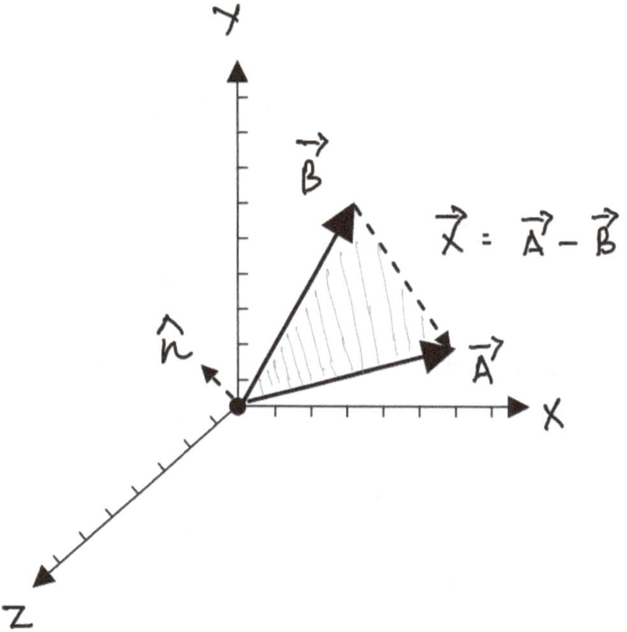

Figure 6.1.13

If vector cross product is a reasonable estimate of how two vectors obstruct one another, what exactly is the cross-product measuring? Mathematically, the magnitude of the cross product is the measure of the area of the parallelogram formed by the two vectors $\vec{A}\ and\ \vec{B}$. We have shown the parallelogram in figure 6.1.14.

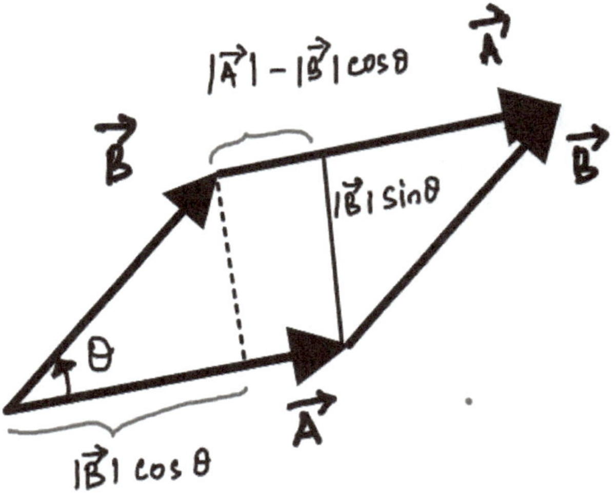

Figure 6.1.14

As depicted in figure 6.1.14, the area of the parallelogram has three parts, i.e., the first triangle, a rectangle, and the third triangle. Both the triangles are of the same area.

$$Area\ of\ the\ parallelogram = 2 \times (Area\ of\ the\ triangle) + Area\ of\ the\ rectangle$$

$$= \left|\vec{B}\right|^2 \cos\theta \sin\theta + \left|\vec{A}\right|\left|\vec{B}\right| \sin\theta - \left|\vec{B}\right|^2 \cos\theta \sin\theta$$

$$= \left|\vec{A}\right|\left|\vec{B}\right| \sin\theta$$

Similar to how we did for the dot product, let us try and get the algebraic equivalent of the cross product. To get there, we shall perform a cross product of two vectors in 3D space.

$$\vec{A} = a_1\vec{\imath} + a_2\vec{\jmath} + a_3\vec{k}$$
$$\vec{B} = b_1\vec{\imath} + b_2\vec{\jmath} + b_3\vec{k}$$

$$\begin{aligned}\vec{A} \times \vec{B} &= \left(a_1\vec{\imath} + a_2\vec{\jmath} + a_3\vec{k}\right) \times \left(b_1\vec{\imath} + b_2\vec{\jmath} + b_3\vec{k}\right)\\ &= a_1b_1\vec{\imath} \times \vec{\imath} + a_1b_2\vec{\imath} \times \vec{\jmath} + a_1b_3\vec{\imath} \times \vec{k} +\\ &\quad a_2b_1\vec{\jmath} \times \vec{\imath} + a_2b_2\vec{\jmath} \times \vec{\jmath} + a_2b_3\vec{\jmath} \times \vec{k} +\\ &\quad a_3b_1\vec{k} \times \vec{\imath} + a_3b_2\vec{k} \times \vec{\jmath} + a_3b_3\vec{k} \times \vec{k}\end{aligned}$$

Equating cross product of the same unit vectors (like $\vec{\imath} \times \vec{\imath}$) to 0 and applying the right-hand rule for the cross products across the unit vectors $\vec{\imath}$, $\vec{\jmath}$ and $\vec{k}$, we shall arrive at the following:

$$\vec{A} \times \vec{B} =$$
$$(a_2b_3 - b_2a_3)\vec{\imath} - (a_1b_3 - b_1a_3)\vec{\jmath} + (a_1b_2 - b_2a_1)\vec{k}$$

A helpful mnemonic to remember the above expression is to organize the two vectors in a 3x3 matrix and calculate its determinant.

$$\vec{A} \times \vec{B} = \begin{vmatrix} \vec{\imath} & \vec{\jmath} & \vec{k} \\ a_1 & a_2 & a_3 \\ b_1 & b_2 & b_3 \end{vmatrix}$$

6.2 Calculus

Here we shall go over the fundamentals of calculus, both differentials, and integrals. Getting a good grasp of calculus is key to fully understanding and deriving the equation $e=mc^2$. Calculus is arguably one of the most misunderstood topics in mathematics. The idea of calculus is intuitive, and in this chapter, I'm hoping to make you feel the intuition behind this simple yet powerful concept. Differential and integrals are two sides of the same coin, and we shall soon see that as well.

As always, our first step to understanding calculus will start with a thought experiment. Let us imagine ourselves to be in a parking lot, about to drive a car. It's a regular car with one crucial difference. The vehicle does not have a speedometer but has a trip meter and a clock. In other words, the dashboard would show the distance you have covered, and the clock would show the elapsed time since the start of the trip. Both the trip meter and the clock can measure both the distance and the time with high precision. Consider driving that car on an empty freeway for 60 minutes. Let us assume that in 60 minutes, we were able to accelerate to a top speed and then decelerate back towards zero. Upon completing our journey, how would the two readings look when plotted against one another on a cartesian coordinate system? A possible depiction is provided in figure 6.2.1. In this figure, we have plotted time on the *x-axis* and distance on the *y-axis*.

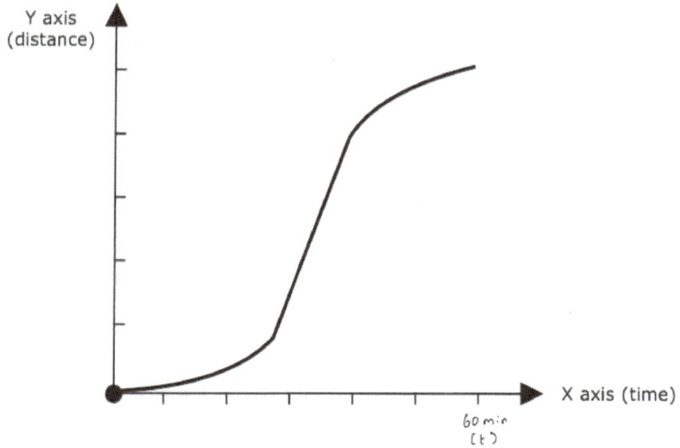

Figure 6.2.1

Since the car did not have a speedometer, what if we pursue to derive the speed (or velocity) in which we traveled at a point in time (say, 30th minute) in our journey? How would we derive that? If, at the end of our journey, our trip meter read 100 miles, could we say our speed is 100 miles per hour? While 100 mph can serve as a crude approximation, it is not the speed at every particular point in our travel time. Remember, we had gradually accelerated before decelerating to a complete stop.

We know that speed is the distance traveled over time. To be precise, let us slice a minute, the sixty seconds that exist between 29th (x_1) and the 30th (x_2) minute, and calculate the difference between the readings on our trip meter, as shown in figure 6.2.2.

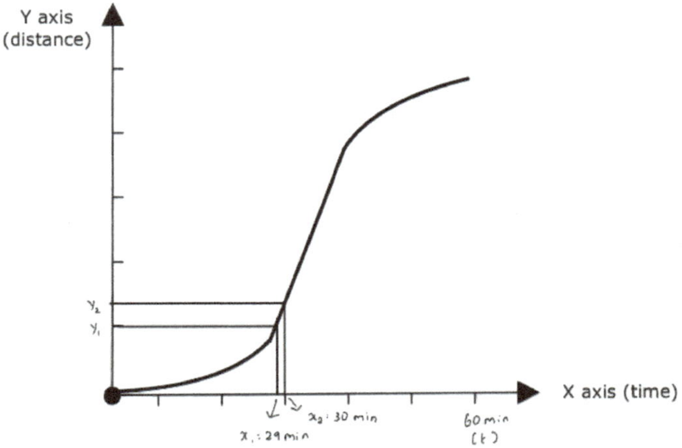

Figure 6.2.2

Now, we have a better approximation of our speed in the 30th minute.

$$Speed\,of\,the\,car\,at\,30th\,min = \frac{y_2 - y_1}{x_2 - x_1}$$

We could make the above approximation even better if we do the calculation between the 29th minute 59 seconds and the 30th minute 0 seconds, i.e., shrink the time slice even smaller. Mathematically, we could get the exact speed if we slice the time to the smallest possible value (close to 0) at the 30th minute and apply the above formula. As shown in figure 6.2.3, if we call that tiny split second as *dx*, and the corresponding difference in distance traveled as *dy*, at any point in time *t*, the speed of the car could be generalized as the following:

$$\textit{Speed of the car at time } x = \frac{dy}{dx}$$

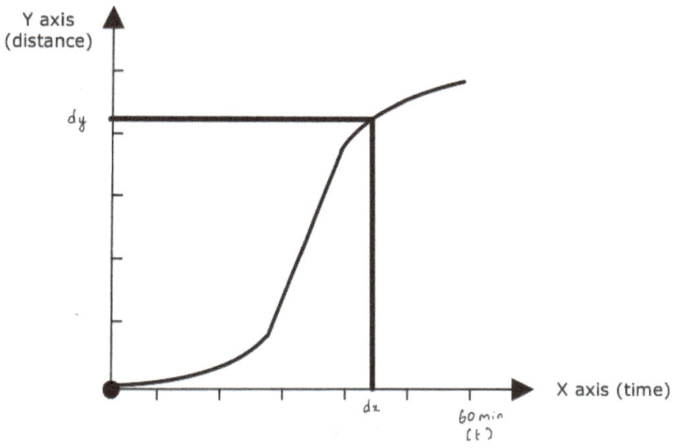

Figure 6.2.3

From the above figure, we can see that the value of *y* depends on x. In other words, *y* is a function of variable *x* or simply *y=f(x)*. So, another way of stating the above equation would be the following:

$$\textit{Speed of the car at time } x = \frac{dy}{dx} = \frac{d}{dx}f(x) = \lim_{\Delta x \to 0} \frac{f(x+\Delta x) - f(x)}{\Delta x}$$

What exactly is *dy/dx*? If you zoom in to find out what is happening at the point in figure 6.2.3, you would see that *dy/dx* is the geometric equivalent of the slope of the tangent line that touches the curve at *x*. The curve is *y* plotted against *x* as depicted in figure 6.2.4. The value *dy/dx* is called the derivative of *y* (or *f(x)*) at *t* and describes the rate of change of

y (or $f(x)$) at x. In our examples, the rate of change of y (or $f(x)$) at time x is the car's speed at time x. So, to derive the car's speed, we would need to differentiate the car's position, y (or $f(x)$), at time x. The process of calculating the derivative of a function is called *differentiation.*

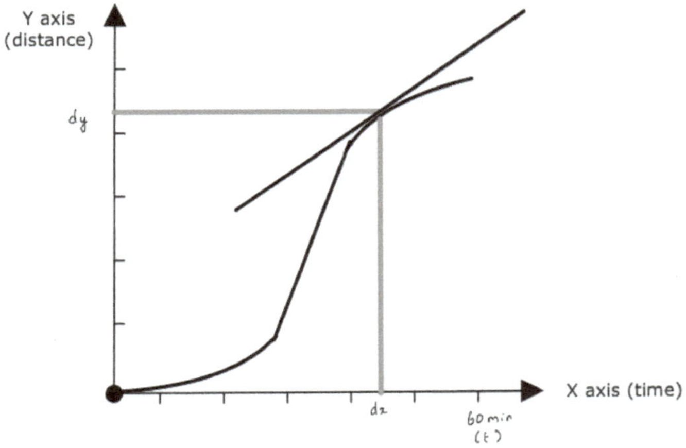

Figure 6.2.4

We can differentiate any quantity that we can express as a function of one or more variables with respect to those variables. We call the outcome of that differentiation the derivative of the original quantity. Here is the crux of differential calculus: *The study of how quantities, expressed as a function of a set of variables, change with respect to their variables.*

So far, we have seen what derivatives are and how we could calculate them. Let us derive one, and for that, we shall consider the following equation. We have also plotted this equation in figure 6.2.5:

$$y = f(x) = x^2$$

$$\frac{dy}{dx} = \frac{d}{dx}f(x) = \lim_{\Delta x \to 0} \frac{f(x + \Delta x) - f(x)}{\Delta x}$$

$$\frac{dy}{dx} = \frac{(x + \Delta x)^2 - x^2}{\Delta x}$$

$$\frac{dy}{dx} = \frac{x^2 + (\Delta x)^2 + 2x\Delta x - x^2}{\Delta x}$$

$$\frac{dy}{dx} = 2x + \Delta x$$

$$\frac{dy}{dx} = 2x, \, when \, \Delta x \longrightarrow 0$$

So, the derivative of $f(x)=x^2$ over x is $2x$. Through the above method, we could also calculate the derivative of x^3 to be $3x^2$. We can generalize the derivative of $f(x)=x^n$ to be nx^{n-1}.

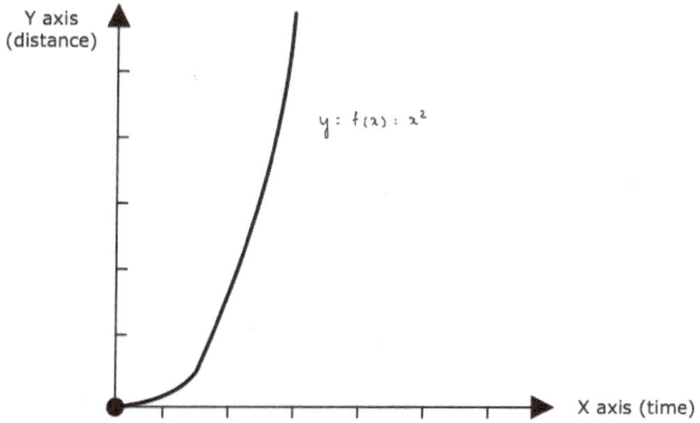

Figure 6.2.5

What about derivatives of constants over variable x? Let us say we have an equation $y=5$. Now, what would dy/dx be? We learned that *dy/dx* denotes the rate of change of *y* at *x*, and for the equation *y=5,* the value of *y* remains the same, i.e., 5 for all values of *x*. So, *dy/dx* is 0 for the equation *y=5*. In other words, the derivative of any constant is 0.

Let us perform our experiment with a different car now. Again, this car is like any other regular car with a slight twist. This car has a high-precision speedometer and a clock but not a trip meter. If we repeat our trip in the same manner, how would the two readings look when plotted against one another on a cartesian coordinate system with time on the *x-axis* and the speed on the *y-axis?* It's likely to be a parabola, as depicted in figure 6.2.6.

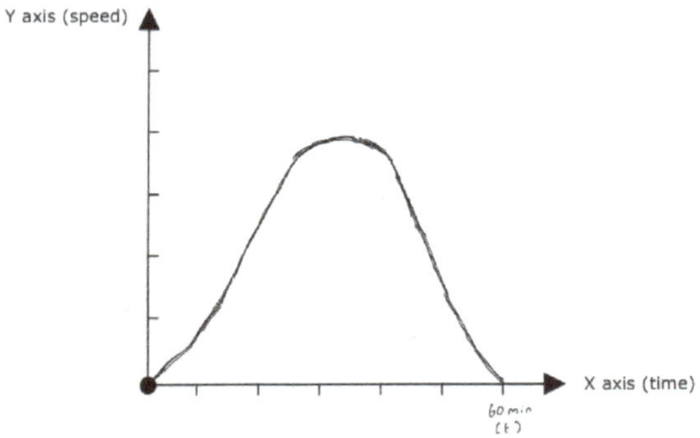

Figure 6.2.6

Since the car does not have a trip meter, how would we determine the distance traveled if we had the above graph? If you notice, the value of *y* at a particular point in the time *x* is the car's speed at time *x*. Since the clock and the speedometer are both of high precision, if we can add up the values of y at every point in time x between 0 and 60 minutes, that should give us the trip's distance. That summed up value is nothing but the area under the curve. So, if we find a way to calculate the area under the curve, we should have the trip's distance. We know how to find the area of any quadrilateral, but for a parabola like the above? We will soon see how the intuition of integrals will help calculate the area under any curve.

To begin with, let us call the equation of the parabola to be $f(x)$, i.e., $y=f(x)$. In figure 6.2.7, you might observe that we could approximate the area under the curve by splitting the curve into rectangles that are a few minutes wide in the *x-axis* and then summing up the individual areas. The width of each rectangle is 10 minutes (on the *x-axis*), and the height is the car's speed at the end of the tenth minute. Since the speed is 0 at the end of the 60th minute, there is no sixth rectangle.

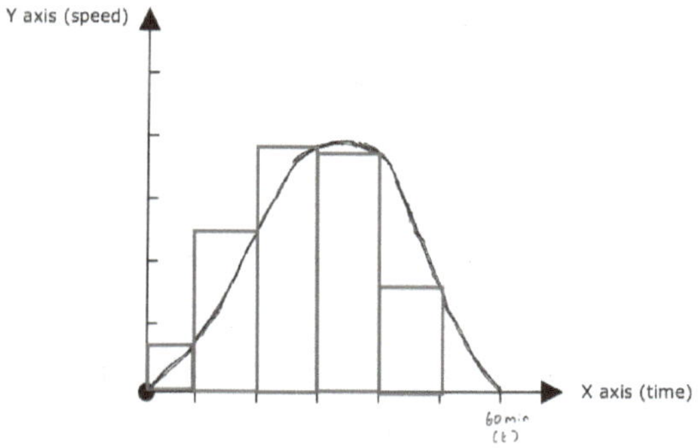

Figure 6.2.7

As depicted in figure 6.2.8, the approximation gets better as we increase the number of splits. If we continue applying this method and splitting the base into n tiny parts, with n approaching *infinity*, our approximation will become accurate, correct? In other words, if we split the duration on the *x-axis* into n infinitesimal parts and then calculate the integrated sum of areas of all those thin slices of rectangles, we should get the accurate area under the curve. While calculating the derivative, we had equated that infinitesimal part to *dx*. We shall make use of that here.

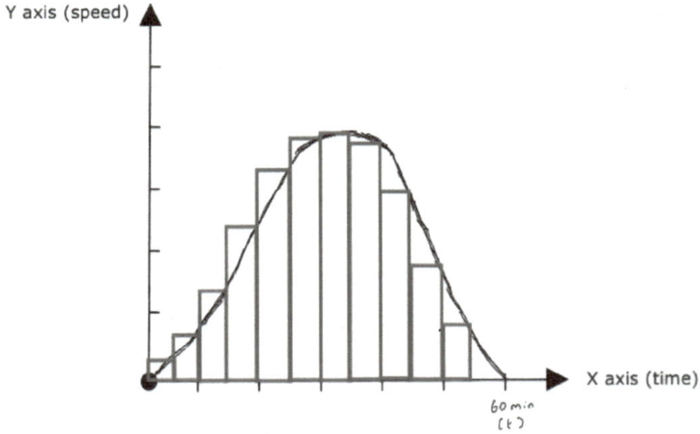

Figure 6.2.8

$$Trip\,distance\,in\,time\,t = Area\,under\,the\,curve = \sum_{x=0}^{x=t} f(x)dx$$

The integrated sum calculated by splitting the area into infinitesimal parts is the integral of the original function *f(x)* over *x*. The process of calculating this integrated sum is called *integration*. We can represent integrals of a function the following way:

$$Trip\,distance\,in\,time\,t = Area\,under\,the\,curve = \int_{x=0}^{x=t} f(x)dx$$

To be precise, what we have stated above is a *definite integral.* The word definite indicates the range involved, i.e., the area under the curve is to be calculated between *x=0* and *x=t.* A definite integral of a function, when integrated over a range, will evaluate to a number. When the same integral is generalized and is made independent of the range, it takes the name of an *indefinite integral.* We can express an indefinite integral the following way:

$$\int ydx$$

Like we did for differentiation, we shall perform integration in flesh and blood. For a specific function, we shall derive both its definite integral and indefinite integral. Our chosen function would be *y=2x.*

$$y = f(x) = 2x$$

$$\int_{x=0}^{x=t} ydx = \int_{x=0}^{x=t} 2xdx = \sum_{x=0}^{x=t} 2xdx$$

$$dx = \frac{t}{n}\ with\, n \longrightarrow \infty$$

$$Values\, of\, x\, would\, range\, from\, 0, \frac{t}{n}, \frac{2t}{n}, \frac{3t}{n} \ldots t$$

$$Values\, of\, y\, would\, range\, from\, 0, \frac{2t}{n}, \frac{4t}{n}, \frac{6t}{n} \ldots 2t$$

$$\int_{x=0}^{x=t} ydx = 0\left(\frac{t}{n}\right) + \left(\frac{2t}{n}\right)\frac{t}{n} + \left(\frac{4t}{n}\right)\frac{t}{n} + \left(\frac{6t}{n}\right)\frac{t}{n} \ldots (2t)\frac{t}{n}$$

$$\int_{x=0}^{x=t} ydx = \frac{2t^2}{n^2}(1 + 2 + 3 + \ldots n)$$

$$\int_{x=0}^{x=t} ydx = \frac{2t^2}{n^2}\left(\frac{n^2+n}{2}\right)$$

$$\int_{x=0}^{x=t} ydx = t^2 + \frac{t^2}{n}$$

$$\int_{x=0}^{x=t} ydx = t^2,\ when\, n \longrightarrow \infty\, then\, \frac{t^2}{n}\, becomes\, 0$$

While t^2 is the definite integral of $y=2x$, when integrated over x between 0 and t, here is the indefinite integral of the same:

$$\int ydx \;= x^2$$

Through the content shared in the last few pages, I hope to have established the intuition of differential and integral calculus. At the beginning of the chapter, we said that differential calculus and integral calculus are two sides of the same coin. How do we explain that? When we learned about differential calculus, we plotted the car's distance over time and derived the speed at time t through differentiation. Later, by integrating speed over time, we could get back to the distance covered in time t. We were able to confirm this when the indefinite integral of function $y=2x$ (integrated over x) yielded the original function ($y=x^2$) that we had used to differentiate. Function A ($f(x)=x^2$) is the antiderivative of another function $B(f(x)=2x)$, if the integral of function B is function A.

Earlier, we had calculated that for a function $y=x^2$, its derivative with respect to x is $2x$. In other words, for every infinitesimal increase in x, the slope of the tangent line drawn at *(x,y)* increases by two times the increase of *x*. As you observe in the below table (figure 6.2.8A), for *x=2*, when *dx* approaches 0, the value of the slope approaches *2x*, i.e., 4.

initial x	initial y	incremental x (dx)	new x	new y	incremental y (dy)	dy/dx
2	4	1	3	9	5	5
2	4	0.75	2.75	7.56	3.56	4.75
2	4	0.5	2.5	6.25	2.25	4.5
2	4	0.25	2.25	5.06	1.06	4.25
2	4	0.1	2.1	4.41	0.41	4.1

Figure 6.2.8A

So far, we have been dealing with functions with only one variable, i.e., *x*. While defining derivatives, we had stated that we could differentiate any quantity that is a function of one or more variables. Going forward, we will try to differentiate functions that have multiple variables. The first function that we shall consider is $f(x,y) = x^2 + y^2$. In this function, both *x* and *y* are variables.

How would we find its derivative? Since the derivative is with respect to one variable, a function with multiple variables can have multiple derivatives. Each one of them is called partial derivatives of the function *f(x,y)*. Our function *f(x,y)* will thus have the following two partial derivatives:

$$\frac{\partial}{\partial x} f(x,y) = 2x + 0 = 2x$$
$$\frac{\partial}{\partial y} f(x,y) = 0 + 2y = 2y$$

As you might observe, while calculating the derivative with respect to (wrt) variable *x*, *y* will be considered a constant. While calculating the derivative wrt variable *y*, *x* will be a constant. Yes, partial derivatives have a slightly different notation, i.e., we denote them by ∂ instead of *d*. Partial derivatives give us a sense of the rate of change of the function wrt one variable. Wouldn't it also be nice to have a way to capture the rate of change of a function with all its variables? In other words, could we calculate the *full derivative* for a function? Yes, there is a way to calculate the same, and it is called the *gradient* (not full derivative). We can express the gradient of a function as $\nabla f(x,y)$.

$$Gradient\, of\, f(x,y) = \nabla f(x,y) = \begin{bmatrix} \frac{\partial}{\partial x} \\ \frac{\partial}{\partial y} \end{bmatrix} f(x,y) = \begin{bmatrix} \frac{\partial}{\partial x} f(x,y) \\ \frac{\partial}{\partial y} f(x,y) \end{bmatrix} = \begin{bmatrix} 2x \\ 2y \end{bmatrix}$$

As you might notice above, the gradient of a function is an organization of all its partial derivatives in the vector form.

$$The\ symbol\ \nabla\ is\ a\ shorthand\ for\ the\ operator \begin{bmatrix} \frac{\partial}{\partial x_1} \\ \frac{\partial}{\partial x_2} \\ .. \\ \frac{\partial}{\partial x_n} \end{bmatrix} or \begin{bmatrix} \frac{\partial}{\partial x_1} & \frac{\partial}{\partial x_2} .. & \frac{\partial}{\partial x_n} \end{bmatrix} across\ n\ variables$$

Why does the gradient have to be a vector? Why can't the gradient also be a scalar, i.e., could we not simply add all the partial derivatives or multiply the partial derivatives to arrive at the *full derivative* i.e., *gradient*? What motivates us to represent a gradient as a vector? We will answer that question by dissecting two key aspects of the gradient. We shall perform that dissection by differentiating (yes, partially) a different function $f(x,y) = 4x^2+y^2$. Our new function $f(x,y)$ has the following partial derivatives:

$$\frac{\partial}{\partial x}f(x,y) = 8x$$

$$\frac{\partial}{\partial y}f(x,y) = 2y$$

You might observe that the function's rate of change with respect to a unit change in x (keeping y constant) is greater than the function's rate of change with respect to a unit change in y (keeping x constant). In other words, the variable x has a more significant influence on the function when compared to y. Here is a demonstration of the same:

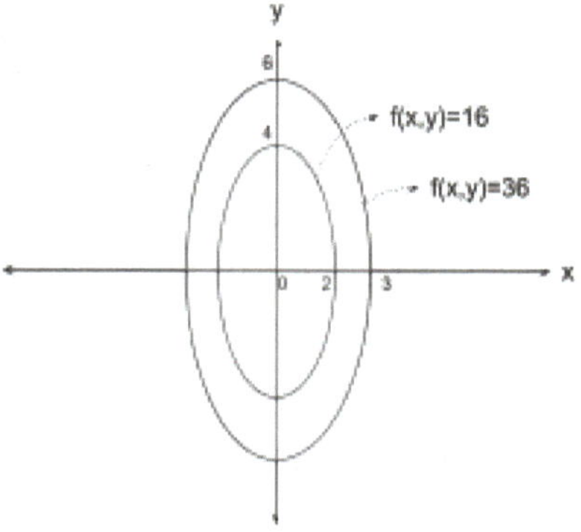

Figure 6.2.9

The two ellipses above are two parabolas viewed from the *z-axis*. The function $f(x,y) = 4x^2+y^2$ defines the two parabola. The inner ellipse corresponds to $f(x,y) = 16$, and the outer ellipse corresponds to $f(x,y) = 36$. Think about the two parabolas as two bowls of varying size and height, when one is placed inside the other and viewed from the top. As you might see in figure 6.2.9, for the function to increase from 16 to 36, x must increase 1 unit (from 2 to 3), while y has to increase two units (from 4 to 6). The first aspect of the gradient is the relative influence each variable has on the function. However, we choose to represent the gradient of a function, it should capture the relative influence of its variables.

While calculating the derivative of a function $f(x)$, i.e., a function with a single variable, we realized that the derivative is equivalent to the slope of the tangent line that touches the curve at point x. The curve is the function $f(x)$ plotted against x. We have seen this in figure 6.2.4. There can be only one such tangent that we can draw at any point in a curve. Since the gradient is informally the *full derivative* of a function with multiple variables, could we say its geometric equivalent as the slope of a tangential line that touches the surface in the multi-dimensional space? Again, the surface here is $f(x,y)$ plotted against x *and* y. A view of the surface formed by $f(x,y) = 4x^2+y^2$ and viewed from the top, i.e., *z-axis,* was provided in figure 6.2.9. Unlike what we saw in figure 6.2.4, we could draw multiple such tangents at a particular point on the multi-dimensional surface. We have articulated this in figure 6.2.10, showing three possible tangents drawn at the same point. Among these various tangents, we define the gradient as a tangent drawn in the direction of the greatest rate (denoted by V) of increasing *f(x,y)*. The second key aspect of the gradient is direction.

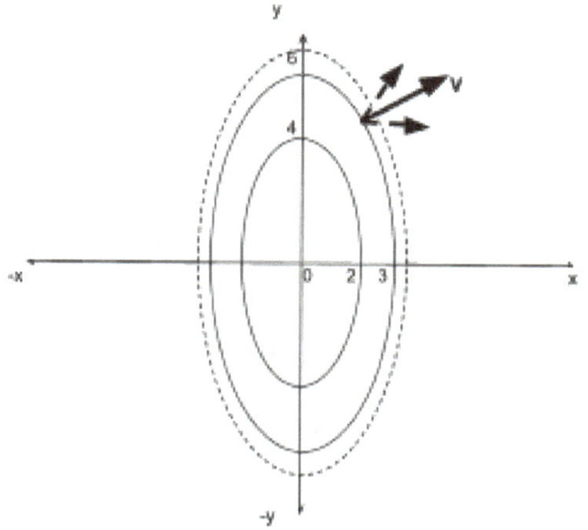

Figure 6.2.10

Representing the gradient as a vector allows us to capture both the above aspects: (1) relative influence of the multiple variables and (2) direction of the greatest rate of increase. Through an example, we shall unravel both these aspects from the gradient vector of the function $f(x,y) = 4x^2+y^2$.

To recap, the gradient of $f(x,y) = 4x^2+y^2$ is given below:

$$Gradient\, of\; f(x,y) = \nabla f(x,y) = \begin{bmatrix} P(x,y) \\ Q(x,y) \end{bmatrix} = \begin{bmatrix} 8x \\ 2y \end{bmatrix}$$

Applying the above formula, here are the precise gradient vector values at four different points.

$$Gradient\,at\,<0,6> \;=\vec{A}=\begin{bmatrix}0\\12\end{bmatrix}=0\vec{\imath}+12\vec{\jmath}$$

$$Gradient\,at\,<1,5.7> \;=\vec{B}=\begin{bmatrix}8\\11.3\end{bmatrix}=8\vec{\imath}+11.3\vec{\jmath}$$

$$Gradient\,at\,<2,4.5> \;=\vec{C}=\begin{bmatrix}16\\8.9\end{bmatrix}=16\vec{\imath}+8.9\vec{\jmath}$$

$$Gradient\,at\,<3,0> \;=\vec{D}=\begin{bmatrix}24\\0\end{bmatrix}=24\vec{\imath}+0\vec{\jmath}$$

In figure 6.2.11, the above gradient vectors are mapped at their respective points on the multi-dimensional surface (*f(x,y)* plotted against *x* and *y*) as well as on a vector space. You could see the direction of each of the four vectors to be in the direction of the greatest rate of increase.

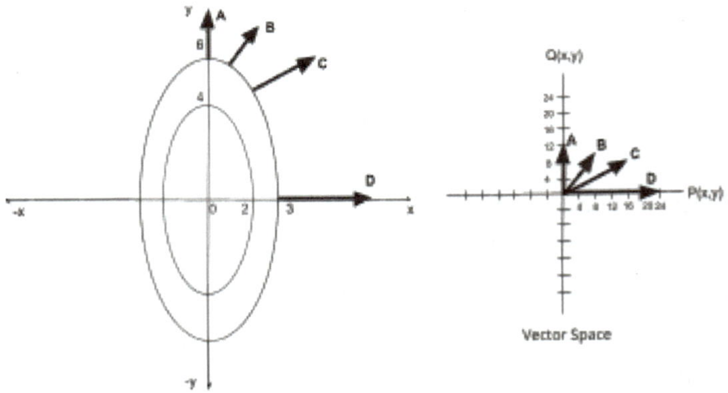

Figure 6.2.11

We can observe the lengths of the gradient vectors increase as we move away from 0 along the *x-axis* and decrease as we move away from 0 along the *y* axis. This indicates the relative influence of the two variables.

$$\vec{A} = 0\vec{i} + 12\vec{j} \quad and \quad \left|\vec{A}\right| = \sqrt{144} = 12$$

$$\vec{B} = 8\vec{i} + 11.3\vec{j} \quad and \quad \left|\vec{B}\right| = \sqrt{192} = 13.9$$

$$\vec{C} = 16\vec{i} + 8.9\vec{j} \quad and \quad \left|\vec{C}\right| = \sqrt{366} = 18.3$$

$$\vec{D} = 24\vec{i} + 0\vec{j} \quad and \quad \left|\vec{D}\right| = \sqrt{576} = 24$$

Now we are ready to learn about two important mathematical ideas that will come in handy when we try to model and understand real-world phenomena through vectors. The real-world concepts could be pressure, electric field, magnetic field, water flow, etc.

As we unravel the intuition behind these ideas, we will also explore their utility values in detail. The two ideas are *divergence* and *curl*. Let us start with *divergence*. Let us consider functions *f(x,y) and f'(x,y)* and their corresponding gradient vectors *G and G'*.

$$f(x, y) = 4x^2 + y^2$$
$$f'(x, y) = -4x^2 - y^2$$

$$Gradient\,of\ f(x, y) = \nabla f(x, y) = \vec{G} = \begin{bmatrix} 8x \\ 2y \end{bmatrix}$$

$$Gradient\,of\ f'(x, y) = \nabla f'(x, y) = \vec{G'} = \begin{bmatrix} -8x \\ -2y \end{bmatrix}$$

Applying the formula for gradient, we shall calculate the precise gradient vector values for *f'(x,y)* at the four points we considered for *G*.

$$\textit{Gradient at} < 0,6 > \; = \vec{A} = \begin{bmatrix} 0 \\ -12 \end{bmatrix} = 0\vec{i} - 12\vec{j}$$

$$\textit{Gradient at} < 1,5.7 > \; = \vec{B} = \begin{bmatrix} -8 \\ -11.3 \end{bmatrix} = -8\vec{i} - 11.3\vec{j}$$

$$\textit{Gradient at} < 2,4.5 > \; = \vec{C} = \begin{bmatrix} -16 \\ -8.9 \end{bmatrix} = -16\vec{i} - 8.9\vec{j}$$

$$\textit{Gradient at} < 3,0 > \; = \vec{D} = \begin{bmatrix} -24 \\ 0 \end{bmatrix} = -24\vec{i} + 0\vec{j}$$

Similar to figure 6.2.11, in figure 6.2.11a, the above gradient vectors are mapped at their respective points on the multi-dimensional surface (*f'(x,y)* plotted against *x* and *y*) as well as on a vector space. Notice the difference in the direction of the gradient vectors between *f(x,y) and f'(x,y)*.

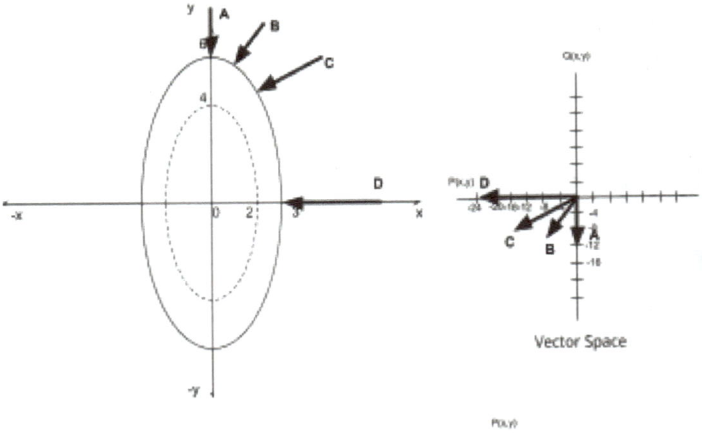

Figure 6.2.11a

In figure 6.2.12, we have captured the general behavior of the gradient vectors for the two functions.

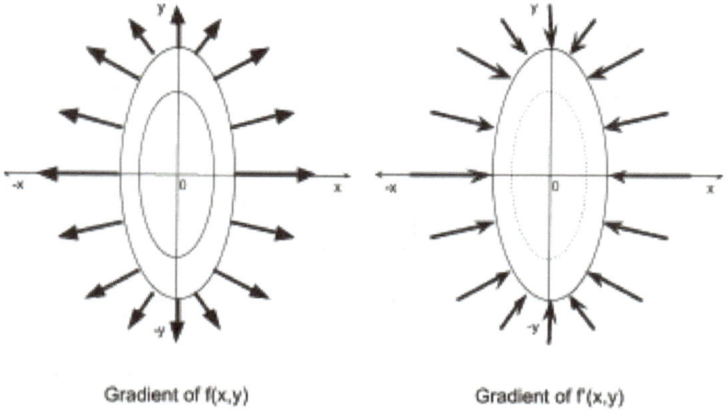

Figure 6.2.12

As you might observe above, vector *G* diverges, i.e., an increase in *x* and *y* increases *f(x,y)*. Similarly, vector *G'* converges, i.e., an increase in *x* and *y* decreases *f(x,y)*. Wouldn't it be effective to derive this aspect through a mathematical calculation instead of drawing and finding out if a vector is converging or diverging? In other words, could we transform any vector of the form *f(x,y)* into a number, i.e., a scalar value that can tell us the degree of divergence or convergence? We will see the utility value of what we would use it for, but for now, let us deliberate on the actual transformation that can help us derive that.

Let us try and identify that transformation logic through patterns in data. In the table below (figure 6.2.13), we have captured the actual values of the Gradient vectors *G* and *G'* for different *x* and *y*. You might see a pattern emerge in them, which we've highlighted in grey. For a vector that converges *(G)*, you could see the values of both *8x* and *2y* to increase in line with an increase in *x* and *y*, respectively. On the other hand, with a diverging vector *G'*, you could see the values of both *-8x* and *-2y* decrease with an increase in *x* and *y*.

Coordinates		Gradient G		Gradient G'	
x	y	8x	2y	-8x	-2y
-3	0.0	-24.0	0.0	24.0	0.0
-2	4.5	-16.0	9.0	16.0	-9.0
-1	5.7	-8.0	11.4	8.0	-11.4
0	6.0	0.0	12.0	0.0	-12.0
1	5.7	8.0	11.4	-8.0	-11.4
2	4.5	16.0	9.0	-16.0	-9.0
3	0.0	24.0	0.0	-24.0	0.0
2	-4.5	16.0	-9.0	-16.0	9.0
1	-5.7	8.0	-11.4	-8.0	11.4
0	-6.0	0.0	-12.0	0.0	12.0
-1	-5.7	-8.0	-11.4	8.0	11.4
-2	-4.5	-16.0	-9.0	16.0	9.0

Figure 6.2.13

How could we capture the above pattern as a formula for any vector represented the following way?

$$f(x, y) = \begin{bmatrix} P(x, y) \\ Q(x, y) \end{bmatrix}$$

We have come to a point to formulate *divergence*. We define *divergence* as the degree of *outflowing-ness* of a vector field:

$$Divergence\, of\; f(x,y) = \frac{\partial}{\partial x}P(x,y) + \frac{\partial}{\partial y}Q(x,y)$$

If you recall the concepts we had read as part of vectors, we could rephrase the above expression as the dot product of ∇ and *f(x,y)*.

$$\begin{aligned} Divergence\, of\; f(x,y) &= \begin{bmatrix} \frac{\partial}{\partial x} \\ \frac{\partial}{\partial y} \end{bmatrix} \cdot \begin{bmatrix} P(x,y) \\ Q(x,y) \end{bmatrix} \\ &= \nabla \cdot f(x,y) \qquad (6.2.1) \end{aligned}$$

In line with the pattern we observed, the two partial derivatives, i.e., *P(x,y)* wrt *x* and *Q(x,y)* wrt *y*, will be positive for a diverging vector. To summarize, the divergence of a vector is a scalar value, and when it is positive, it means the vector is diverging, and when it is negative, it means the vector is converging. The actual value shall signify the strength of divergence or convergence.

Okay, we have accomplished what we intended to, i.e., to arrive at a convenient scalar value to calculate the degree of convergence or divergence. What about its utility value? Let us take a real-life scenario: Consider a vast marshland filled with water in the city's outskirts. The spread of the land is vast in the scale of tens to hundreds of square miles. We know that there are some water sources in the marshland on the ground, and our pursuit is to find their exact location. Let us say we have access to a camera, a drone, and a computer. Could we apply the following approach and solve the problem at hand?

1. Divide the entire area into smaller blocks of squares.
2. Survey each square block with the help of the camera fitted on the drone. We shall have the camera record the movement of water in that block.
3. We could process each of the video clips and transform the movement of the water into a vector.
4. Once we model the water's movement as a vector, we can calculate its divergence value according to the formula in equation 6.2.1.

If done accurately, the square block with positive divergence would lead us to the approximate location of the water source. What about the other square blocks? What would their divergence values be? If the marshland had only one source, then the divergence of the remaining square blocks would not be a positive number. Would the divergence value be negative for them? Unless there is a sink to collect water, that will not be the case either. So, square blocks free of both source (divergence) and a sink (convergence) will have a divergence value of *zero*.

As we wrap up divergence, let us review the divergence of a vector $f(x,y,z)$ in 3D space.

$$f(x,y,z) = \begin{bmatrix} P(x,y,z) \\ Q(x,y,z) \\ R(x,y,z) \end{bmatrix}$$

$$\begin{aligned} Divergence\, of\, f(x,y,z) &= \begin{bmatrix} \frac{\partial}{\partial x} \\ \frac{\partial}{\partial y} \\ \frac{\partial}{\partial z} \end{bmatrix} \cdot \begin{bmatrix} P(x,y,z) \\ Q(x,y,z) \\ R(x,y,z) \end{bmatrix} \\ &= \nabla \cdot f(x,y,z) \qquad (6.2.2) \end{aligned}$$

Our pursuit of identifying the square block with the water source in our marshland may not be as straightforward as we have articulated in the earlier paragraphs. Some blocks that are free of both sources and sinks can have swirls of water too. These can trick our camera and calculations and create an illusion of a water source. So, once we transform the movement of water into a vector, we need to have a way to identify swirls to exclude those square blocks that have them. To do that, we have our second important mathematical idea—*curl.*

In figure 6.2.14, we have shown the example of a vector that swirls around a point in the counterclockwise direction. If you observe, a swirl will have two properties—the velocity (angular velocity to be precise) of the swirl and the orientation of the swirl, i.e., the axis of rotation. Capturing both mathematically as *curl* would imply *curl* has to be a vector that is normal to the plane of the swirl and whose magnitude equates to the angular velocity.

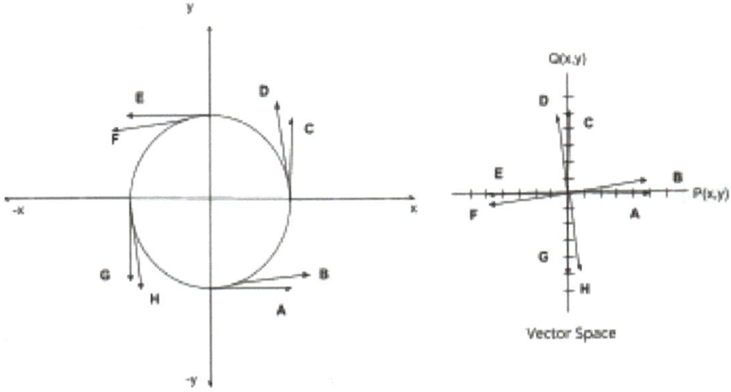

Figure 6.2.14

Like we did for divergence, we shall try and formulate the mathematical formula for the *curl* of a vector *f(x,y,z)*.

$$f(x, y, z) = \begin{bmatrix} P(x, y, z) \\ Q(x, y, z) \\ R(x, y, z) \end{bmatrix}$$

Before we get to measure the angular velocity on a 3D plane, let us see how we would measure the angular velocity in 2D space, as shown in figure 6.2.14. For this calculation, we shall consider R*(x,y,z)* to be zero. Observing the behavior of our vector in the 2D space (left of figure 6.2.14) and in its vector space (right of figure 6.2.14), we could see the following pattern emerge:

- An increase in the value of *x* always increases the value of *Q(x,y,z)*. It would mean the following:
$$\frac{\partial}{\partial x} Q(x, y, z) \; is \; always \; positive$$

- An increase in the value of *y* always decreases the value of *P(x,y,z)*. It would mean the following:
$$\frac{\partial}{\partial y}P(x, y, z)\, is\, always\, negative$$

For the above pattern to emerge, the below expression should hold good:

$$\frac{\partial}{\partial x}Q(x, y, z) - \frac{\partial}{\partial y}P(x, y, z) > 0$$

What about the scenario of a clockwise swirl? If you plot a clockwise swirl in a vector space as shown in figure 6.2.14, you would see the following expression to hold good.

$$\frac{\partial}{\partial x}Q(x, y, z) - \frac{\partial}{\partial y}P(x, y, z) < 0 \; or \; \frac{\partial}{\partial y}P(x, y, z) - \frac{\partial}{\partial x}Q(x, y, z) > 0$$

In summary, through the below expression, we have the magnitude of the angular velocity of a swirl with the sign representing the direction of the swirl (counterclockwise swirl represented by positive value and vice versa).

$$\frac{\partial}{\partial x}Q(x, y, z) - \frac{\partial}{\partial y}P(x, y, z)$$

When we represent the above expression as a vector in the direction of the *z-axis*, we have the curl of our vector.

$$\left(\frac{\partial}{\partial x}Q(x, y, z) - \frac{\partial}{\partial y}P(x, y, z)\right)\vec{k}$$

To generalize the above expression for any 3D plane, we need to derive the respective coefficients for $\vec{i}\ and\ \vec{j}$. In other words, the angular velocity from the perspective of the *x-axis* and *y-axis*. The following expression measures the velocity of the same swirl from the perspective of the *x-axis*:

$$\frac{\partial}{\partial y}R(x,y,z) - \frac{\partial}{\partial z}Q(x,y,z)$$

Finally, the following expression measures the velocity of the same swirl from the perspective of the *y* axis. Remember, a swirl that appears counterclockwise from the *z-axis* and *x-axis* will appear clockwise from the *y-axis*. Hence the change in the sign below.

$$\frac{\partial}{\partial z}P(x,y,z) - \frac{\partial}{\partial x}R(x,y,z)$$

We are now ready to define the curl of any vector in the 3-dimensional space:

$$Curl\ of\ f(x,y,z) =$$

$$= \left[\frac{\partial}{\partial y}R(x,y,z) - \frac{\partial}{\partial z}Q(x,y,z), \frac{\partial}{\partial z}P(x,y,z) - \frac{\partial}{\partial x}R(x,y,z), \frac{\partial}{\partial x}Q(x,y,z) - \frac{\partial}{\partial y}P(x,y,z)\right]$$

$$= \left(\frac{\partial}{\partial y}R(x,y,z) - \frac{\partial}{\partial z}Q(x,y,z)\right)\vec{i} + \left(\frac{\partial}{\partial z}P(x,y,z) - \frac{\partial}{\partial x}R(x,y,z)\right)\vec{j} + \left(\frac{\partial}{\partial x}Q(x,y,z) - \frac{\partial}{\partial y}P(x,y,z)\right)\vec{k}$$

In the above equation, $\vec{i}, \vec{j}, \vec{k}$ are unit vectors in the *x*, *y*, and *z-axis*.

$$= \left(\frac{\partial}{\partial y}R(x,y,z) - \frac{\partial}{\partial z}Q(x,y,z)\right)\vec{i} - \left(\frac{\partial}{\partial x}R(x,y,z) - \frac{\partial}{\partial z}P(x,y,z)\right)\vec{j} + \left(\frac{\partial}{\partial x}Q(x,y,z) - \frac{\partial}{\partial y}P(x,y,z)\right)\vec{k}$$

We could apply the same mnemonic that we used for the cross product of two vectors. We could organize the above expression as the determinant of a 3x3 matrix, as shown below.

$$Curl\,of\,f(x,y,z) = \nabla \times f(x,y,z) = \begin{vmatrix} \vec{i} & \vec{j} & \vec{k} \\ \frac{\partial}{\partial x} & \frac{\partial}{\partial y} & \frac{\partial}{\partial z} \\ P(x,y,z) & Q(x,y,z) & R(x,y,z) \end{vmatrix} \quad (6.2.3)$$

Let us come back to the fundamentals of integrals. We arrived at a way to calculate the area under the curve between two points through integration. We defined integration as splitting the curve into infinitesimal parts along the axis where the two points exist and then calculating its integrated sum. The length of that infinitesimal part is quantified as *dx* when we intend to integrate along the *x-axis*. We have shown this in figure 6.2.15.

$$Area\,under\,the\,curve\,between\,a\,and\,b = \int_a^b f(x)dx$$

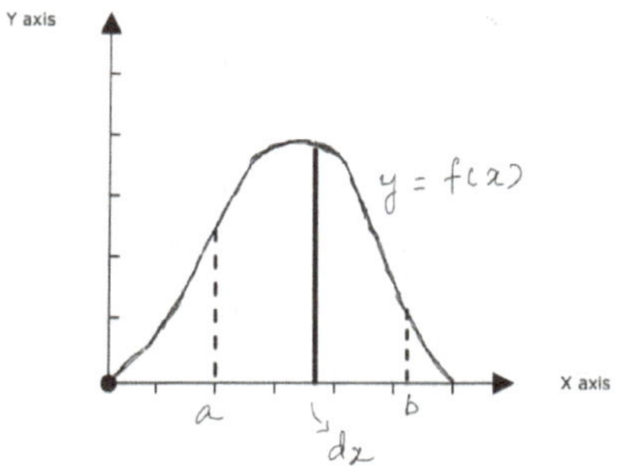

Figure 6.2.15

This notion of splitting something into infinitesimal parts and then calculating its overall sum can be generalized, which we shall discuss in the following pages. We shall explore different ways to generalize the idea of integration. As a first step, we shall not call out the coordinates as shown in figure 6.2.16.

$$Area\, under\, the\, curve\, between\, t_1\, and\, t_2 = \int_{t_1}^{t_2} f(t)dt$$

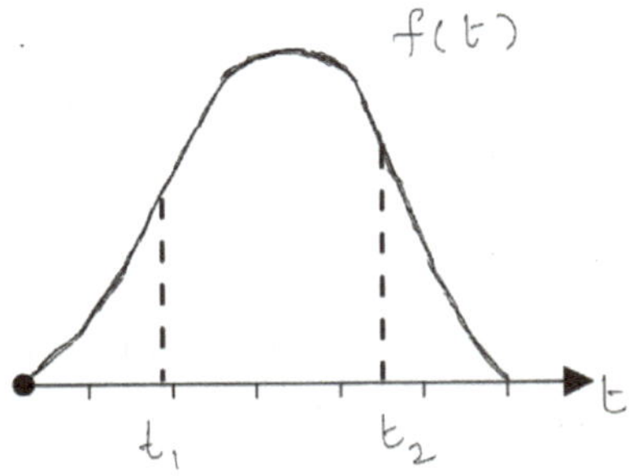

Figure 6.2.16

Not just with scalar values, we can integrate vector quantities too. What exactly is the motivation to integrate vectors? We shall see that through an example. In figure 6.2.17, we have a trolley that we need to push from point A to B. The two points are *d* meters apart. What work would we have accomplished if we applied a force of *n* Newtons in the direction towards B (from A)? In section 2, we defined *work* as the force (experienced by an object) times the displacement of that particular object. More accurately, we define work as the magnitude of force applied in the direction of displacement times the magnitude of displacement. In figure 6.2.17, since the direction of force is in line with the direction of displacement, work accomplished is *nd* joules. To be precise, if we represent Force and the Displacement by vectors $\vec{F}$ *and* $\vec{D}$, then work performed can be derived through its dot product.

$$Work\,done\,W = \vec{F} \cdot \vec{D}$$

Here is the calculation of the work accomplished for the example just considered:

$$\vec{F} = n\vec{i} + 0\vec{j}$$
$$\vec{D} = d\vec{i} + 0\vec{j}$$

$$Work\,done\,W = \vec{F} \cdot \vec{D} = \left|\vec{F}\right|\left|\vec{D}\right| \cos 0 = nd$$

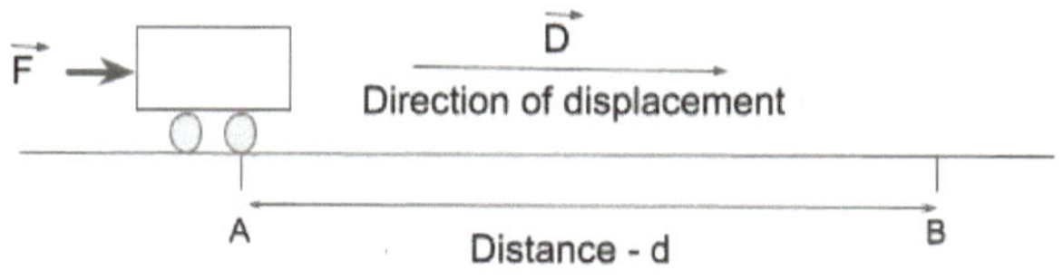

Figure 6.2.17

Here comes the interesting part. What if the path is not a straight line? Let us say we will have to push the trolley on a curved railway track. On a curving track, as shown in figure 6.2.18, if we manage to push the trolley from *A'* with the same amount of force, what would be the work accomplished by the time we reach B? Unlike the earlier diagram, in figure 6.2.18, the view is from the top.

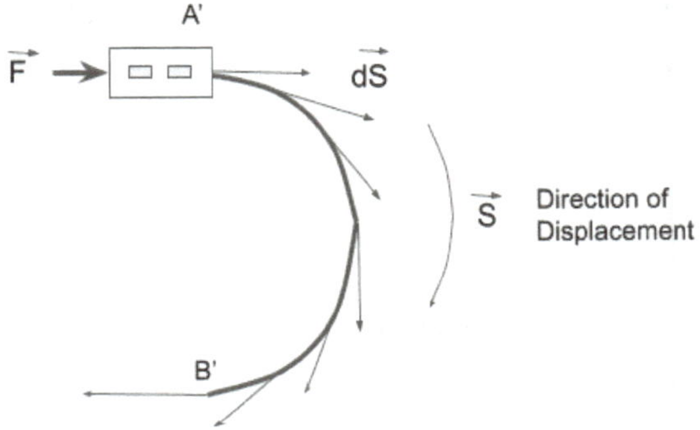

Figure 6.2.18

We saw work's definition as the magnitude of displacement times the magnitude of the work in the direction of displacement. On this curved surface, the direction of displacement changes at every point. At every point, we can represent the direction of displacement by the tangent drawn at that point. If vector $\vec{S}$ represents the tangents (of unit length) drawn at every point on the curve, then we could use our idea of integration to derive the work accomplished in the following way:

$$Work\,done\,W' = \int_C \vec{F} \cdot d\vec{S}$$

In the above equation, the letter C denotes that we are integrating along a curve, and $d\vec{S}$ refers to the unit tangent vector drawn at a particular point in the curve. We can expand the above equation as follows:

$$Work\,done\,W' = \int_C \vec{F} \cdot d\vec{S} = \int_C \vec{F} \cdot \hat{n} ds$$

In the above equation, $\hat{n}$ is a unit tangent vector at an infinitesimal point, i.e., *ds* on the curve. The standard notation is to equate $\hat{n}ds$ as $d\vec{S}$.

Earlier in the chapter, we established integrals by evaluating a function (involving a single variable) along the *x-axis*. Line integrals are a mere generalization of that idea, wherein we evaluate a function along any line or curve. Before we move to the next generalization technique, there is one other point that we would need to cover in line integrals. What if our curve is a closed loop? This particular case is denoted with a circle in the integration symbol, as shown below.

$$\oint_C \vec{F} \cdot d\vec{S}$$

So far, we have been dealing with functions with single variables. Now let us expand to functions that involve multiple variables, and we shall start with two-dimensional space. Let us imagine the representation of the United States map in two-dimensional space as shown in figure 6.2.19. We can understand that the population density would vary across the different counties. What if we have to calculate the total population of America? Suppose we can represent the population density at various points as a function *f(x,y)*. Could we apply our idea of integration here to arrive at the total population of the county? What would we get if we integrate *f(x,y)* first over *x* and then over *y*?

$$\int \left(\int f(x,y)dx \right) dy$$

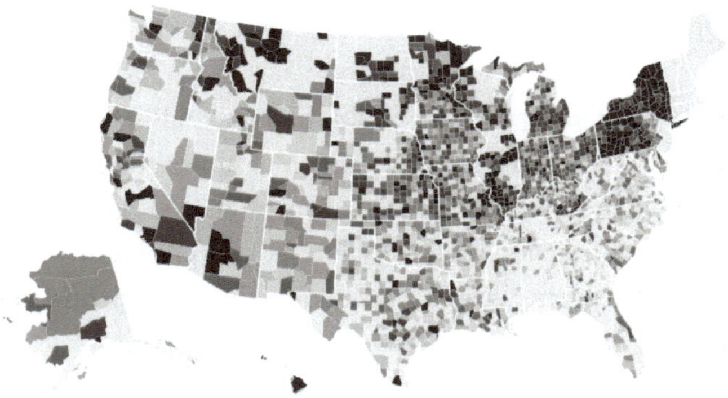

Figure 6.2.19

We have seen how integrating a function with a single variable gets us the area under the curve. Similarly, integrating a function with two variables will provide the volume under the curve (as shown in figure 6.2.20). In our example, through the above double integration on *f(x,y)*, the derived volume under the curve would yield the entire city's population. We could reorganize the previous equation differently, as shown below:

$$Total\ Population = Volume\ under\ the\ curve = \int \left(\int f(x,y)dx \right) dy$$
$$= \int \int f(x,y)dxdy$$
$$= \int \int f(x,y)dA$$

Similar to dx, dA is that infinitesimal rectangle formed in the xy plane. If the notion of integrals (involving one variable) integrates over intervals, then the idea of double integrals integrates over rectangles.

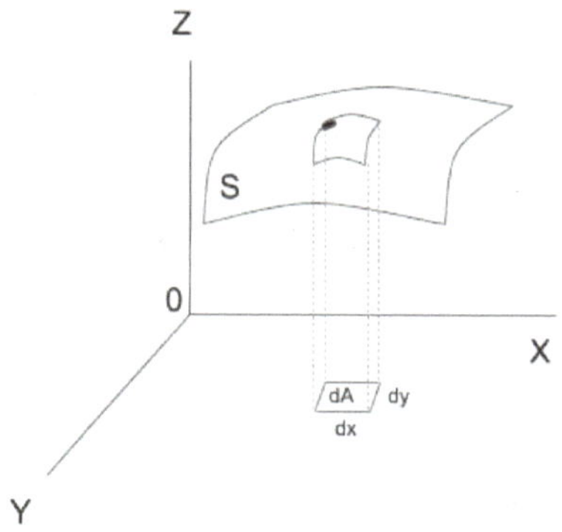

Figure 6.2.20

Sometime back, we explored the idea of line integrals, which was the generalization of integrals (involving one variable). On a similar line, is there a generalization of double integrals? The answer is yes, and the generalization of double integrals is called surface integrals.

Our motivation for line integrals was to measure work done by a force (a vector) along a curved terrain. What motivates us for surface integrals? The answer lies in the measurement of flux. What exactly is flux? Flux represents the rate of flow of something through a given surface. Let us take an example of a garden sprinkler that sprinkles water in all directions, as shown in figure 6.2.21.

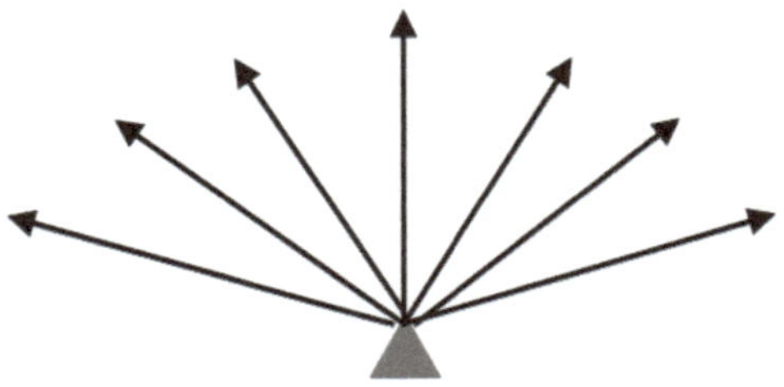

Figure 6.2.21

Now imagine a wired net with minute pores big enough to let water particles flow through it. The net is foldable and consider placing this net in three different ways above the sprinkler, as shown in figure 6.2.22. The water's speed remains the same in all three scenarios. Now, which position of the net is likely to allow more water in unit time? If your answer is B, you are ready to understand what flux is. Position B allows water to flow through the net with the least obstruction. Thus, position B has the highest flux (of the three).

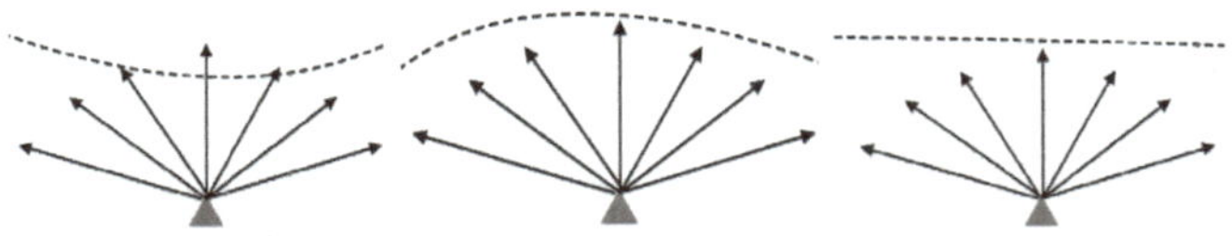

Figure 6.2.22 (Position A, B, and C)

So, how do we measure flux in our example? In the context of our example, we understand flux to be the quantity of the water passing through the surface of the entire net. For a moment, let us keep the formula for calculation of flux aside and see how our idea of integration could help. Remember, the speed of the water as it comes out of the sprinkler is constant. In each of the three positions, if somehow, we can calculate the flux (let us call it F for now) in an infinitesimal part of the surface (let us call it dA), could we integrate that over the entire surface area to derive the total flux? This infinitesimal part on the curved surface will be a curved rectangle in positions A and B but a flat rectangle in position C. This is shown in figure 6.2.23. We just proposed the generalization of a double integral, which is the notion of integrating over any surface (beyond planes). The letter S denotes integration over a surface.

$$Total\ flux = \int_S F dA$$

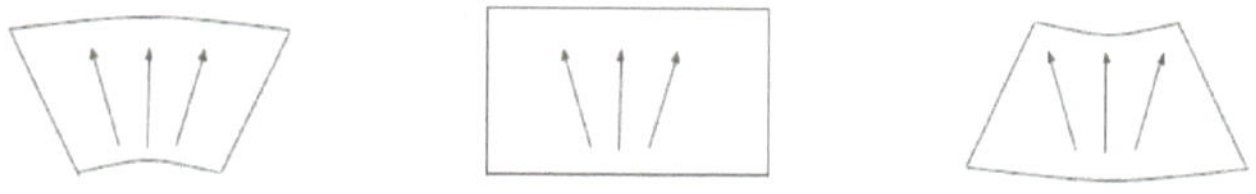

Surface area dA across positions A, B and C

Figure 6.2.23

Now, how would we calculate the flux (F) at that infinitesimal part? We know that flux is high when more water can pass through the net in unit time. If we represent the direction of water flow from the sprinkler by $\vec{W}$ and $\hat{n}$ to be a unit vector that is normal to the surface, then what would the dot product of the two indicate? The dot product will be highest when the angle between them is 0. The dot product will decrease as the angle between $\vec{W}$ and $\hat{n}$ starts to increase. We can see this in figure 6.2.24. We can define total flux as the following.

$$Total\ flux = \int_S \vec{W} \cdot \hat{n} dA$$

As we learned in line integrals, the standard notation is to equate $\hat{n}dA$ to $d\vec{A}$, and restate the previous equation as follows:

$$Total\ flux = \int_S \vec{W} \cdot d\vec{A}$$

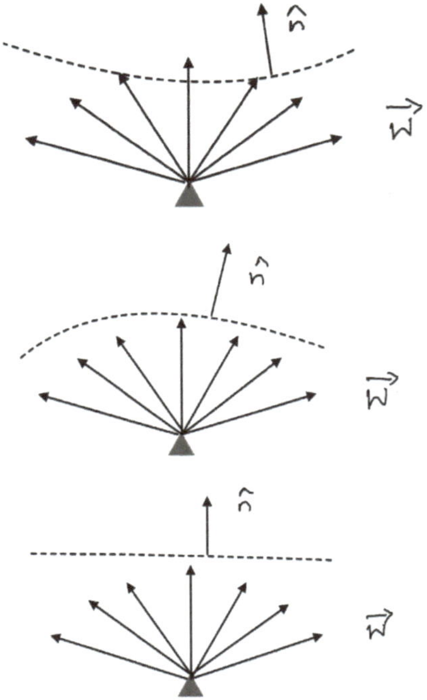

Figure 6.2.24

In our example, the curved surface was not closed. There can be scenarios (like luminous flux) where the surface will be closed. Luminous flux describes a light source's quantity of light emitted across all directions. If you buy a light bulb from a store, you can see its brightness or light quantity described in *lumens*: Higher the lumen value, higher the flux, and brighter the bulb. Like the closed line integral, we denote the closed surface integral with a circle in the integration symbol. Here L represents the luminous flux in an infinitesimal surface part represented by dA.

$$\oint_S LdA$$

The last generalization that we will explore is an extension of the double integration called triple integration. As we indicated, triple integration is extending double integration to three dimensions. Let us imagine a solid cube whose density varies from the center through the edge. How would we measure the mass of the entire cube? If we have a function that would provide the density of a specific speck of the cube, we could integrate it through the volume of the cube and derive its entire sum, as shown below:

$$\begin{aligned} Total\, mass &= \int \left(\int \left(\int f(x,y,z)dx \right) dy \right) dz \\ &= \int \int \int f(x,y,z)dxdydz \\ &= \int \int \int f(x,y,z)dV \end{aligned}$$

Like *dx and dA*, *dV* is that infinitesimal cuboid formed in the *xyz* plane, as shown in figure 6.2.25. Suppose the notion of integrals (involving one variable) integrates over intervals, and the notion of double integrals is to integrate over rectangles. Likewise, the idea of triple integrals integrates over cuboids.

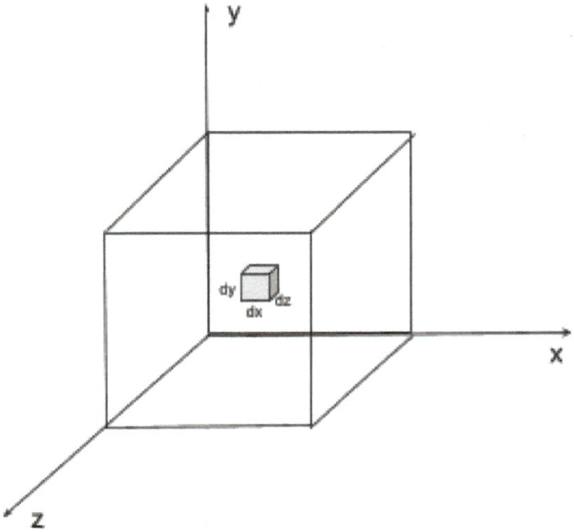

Figure 6.2.25

As you might have guessed by now, the generalization of the triple integral is volume integral. Similar to work (for the line integral) and flux (for surface integral), let us start with a practical example that could motivate us to the relevance of volume integrals. How could we mathematically estimate the total amount of sunlight emitted by the sun? Sun emits sunlight in all directions, and planet earth receives a portion of it. As shown in figure 6.2.26, using the dimensions of the earth and its distance from the sun, let us suppose we can measure the amount of sunlight received by the earth (shown through the three-dimensional cone). We could estimate the total sunlight as an integrated sum across the volume of the three-dimensional sphere.

$$Total\ sunlight\ emitted\ by\ the\ sun = \int_V EdV$$

In the above equation, E corresponds to the amount of sunlight received by the earth. What is received by the earth is a slice of the total sunlight emitted by the sun. We have denoted this slice as a three-dimensional cone whose volume is *dV*. Integrating this over the entire sphere will yield us a reasonable estimate of the total sunlight. Since the sphere is a closed surface, a more accurate notation would represent volume integration with a circle.

$$Total\ sunlight\ emitted\ by\ the\ sun = \oint_V EdV$$

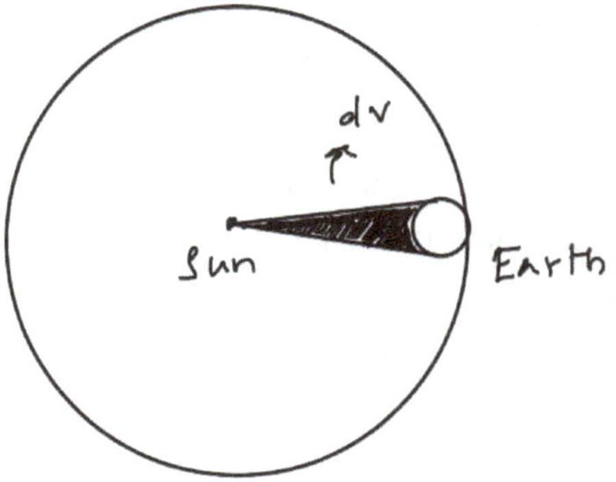

Figure 6.2.26

How would we calculate *E*? If vector $\vec{F}$ represents the sunlight emitted by the sun, its divergence would be a reasonable estimate of E. If you remember, divergence is the degree of outflowing-ness of a vector field, which, in our case, is sunlight.

$$Total\,sunlight\,emitted\,by\,the\,sun = \oint_V \nabla \cdot \vec{F} dV$$

Now that we have seen three different ways to generalize integration, we shall explore their relationships. Let us take the most recent example that we considered in volume integration.

We explored how a measure of sunlight towards the earth can help us estimate the sun's total sunlight through volume integration. In figure 6.2.26, if we take the earth away from the huge sphere surrounding the sun, then the volume of sunlight that passes through the three-dimensional pie will also get past the surface of the sphere, as shown in figure 6.2.27. If we denote the surface area of that portion by *dS* (with *S* denoting the surface area of the entire sphere), then the amount of sunlight that passes through the portion *dS* could be measured through a double integral, which we know is the flux of the sunlight. We know that the flux measured across *dS* will be no different than the amount of sunlight that passes through the three-dimensional pie whose volume is *dV*.

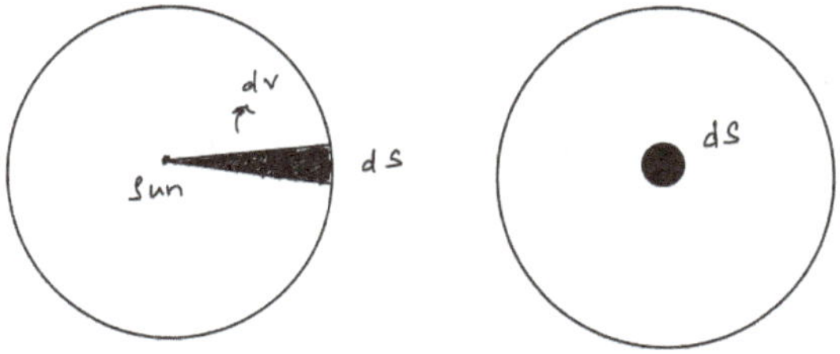

Figure 6.2.27

We are now ready to state the first relation formally: *the surface integral of a vector field over a surface called the flux equals the volume integral of the divergence over the region inside the surface.* We have just stated the *divergence theorem*, also known as Gauss's theorem or Ostrogradsky's theorem. We can mathematically state Gauss's theorem as follows:

$$\int_V \nabla \cdot \vec{F} dV = \int_S \vec{F} \cdot \hat{n} dS$$

As we move to explore the next relation amongst integration, for a moment, let us imagine the formation of its swirl in a tornado.

Image by yasioo from from pixabay

Figure 6.2.28

Figure 6.2.29 captures the path of a particle from the ground lifted by the force of a counterclockwise swirl caused by a tornado. The image on the left shows the skeleton of the path when seen from the ground. The one on the right shows the aerial view of the same swirl. As you might have observed, there will hardly be a gap between swirls in an actual tornado. In the below skeleton, the only intention of showing swirls with gaps is to capture the particle's path.

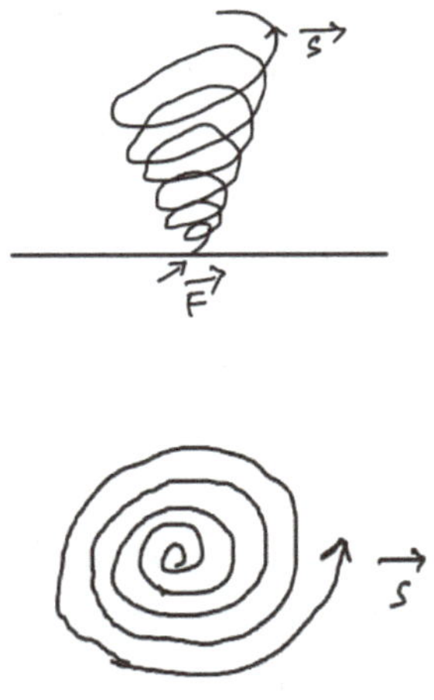

Figure 6.2.29

From the above picture, we can make the following observations:

- A particle starts with an initial force $\vec{F}$ from the ground. A chain of curls influences the particle to take the path of a swirl (represented by $\vec{S}$).
- Though the chain of curls influences the particle, the particle does diverge as it elevates above the ground.

From the above observations, we could calculate the intensity of the tornado in two different ways, and both would be the same:

1. As an integrated sum of $\vec{F}$ along the curved path $\vec{S}$. We can calculate this through the following line integral:

$$\int_C \vec{F} \cdot d\vec{S}$$

2. We can also calculate the intensity as an integrated sum of the flux of the curl experienced by the particle at every point on the surface. Why so? That's because, between two curls, we see the particle diverge, and that tendency to diverge would be felt as flux at every point on the surface of the tornado. When we visualize the tornado as a solid cone, we can represent the flux of the curl throughout its surface as the following surface integral:

$$\int_S \nabla \times \vec{F} \cdot \hat{n}\, dA \ (or) \int_S \nabla \times \vec{F} \cdot d\vec{A}$$

$\hat{n}$ is the unit vector normal to the surface, and dA is an infinitesimal part along the surface of the tornado. The relation between line integrals and surface integrals that we have just arrived at forms the crux of the Stoke's theorem: *The line integral of a vector field over a loop is equal to the flux of its curl through the enclosed surface.* Mathematically it is represented the following way:

$$\int_C \vec{F} \cdot d\vec{S} = \int_S \nabla \times \vec{F} \cdot \hat{n}\, dA$$

6.3 Alternate coordinate systems

So far, we have been using the Cartesian Coordinate system to learn the concepts of vectors and calculus. In this chapter, we shall explore another coordinate system, called the spherical coordinate system. It will come in handy as we begin to apply the concepts we have learned to understand the science behind electromagnetism, especially Maxwell's equations. Let us start with a refresher on the cartesian coordinate system.

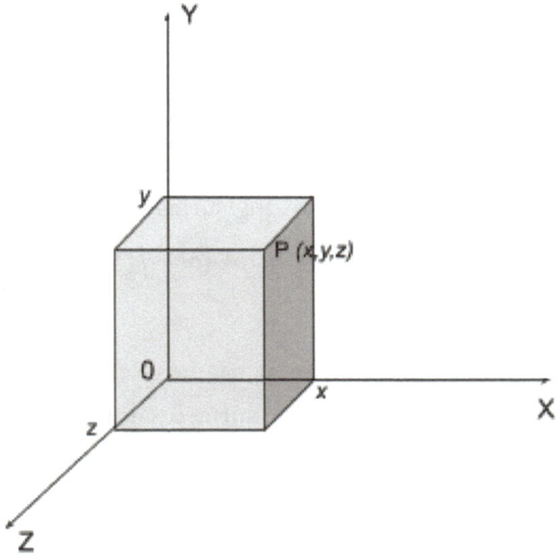

Figure 6.3.1

As you might observe from figure 6.3.1, a three-dimensional cartesian coordinate system has three coordinate axes (*x, y, and z*) pairwise and perpendicular. In addition, the three axes have a single unit of length and intersect at a common point (the origin). A point in space P is referred to by a group of three numbers: (x,y,z). The number x corresponds to the point where a hyperplane, when drawn perpendicular to the *x-axis* and through point P, lands on the *x-axis*. The same applies to *y and z*. We can see this in figure 6.3.1.

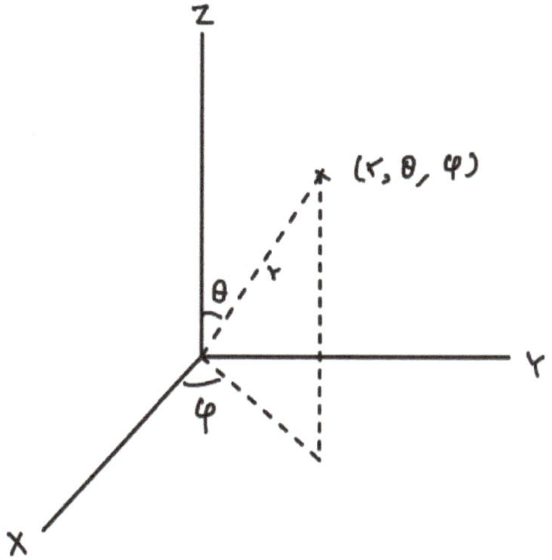

Figure 6.3.2

In the spherical coordinate system (figure 6.3.2), we can refer to the same point using the following triplet:

$$(r, \theta, \phi)$$

- r refers to the radial distance
- θ refers to the polar angle, i.e., angle with respect to the polar axis
- ϕ refers to the azimuthal angle, i.e., angle of rotation from the initial meridian plane.

If the definition sounds complex, don't worry. You may ignore the jargon. The crux of the spherical coordinate system is to refer to a point using a radial distance and two angles, as shown in figure 6.3.2. Another key difference with the spherical coordinate system is that the angles are not in degrees but radians. This aspect is another reason why the spherical coordinate system is a preferred one in physics. Salman Khan (Khan Academy) says there is a degree of purity in the radians, and we shall soon explore how the concept of a radian is so intrinsic to a circle. Also, the idea of radian leads us to measure the angle around a point in 3-dimensional space, which is crucial to understanding electrostatic force that is relevant in our journey to understanding $e=mc^2$.

All of us might know the formula to calculate the circumference of a circle with radius R, which is $2\pi R$ or πD . Here *D is* the length of the diameter of the circle, and *D* equals *2R*. The mathematical constant π captures the relation between the length of the circumference of any circle and its radius:

$$The\, length\, of\, the\, curcumference\, of\, any\, circle\, is\, 6\frac{2}{7} \times length\, of\, its\, radius.$$

$$Circumference\, of\, a\, circle\; = 2\pi R = \pi D$$

As shown in figure 6.3.3, on a two-dimensional plane, we define one radian as the angle subtended at the center of the circle by an arc, whose length equates to the circle's radius.

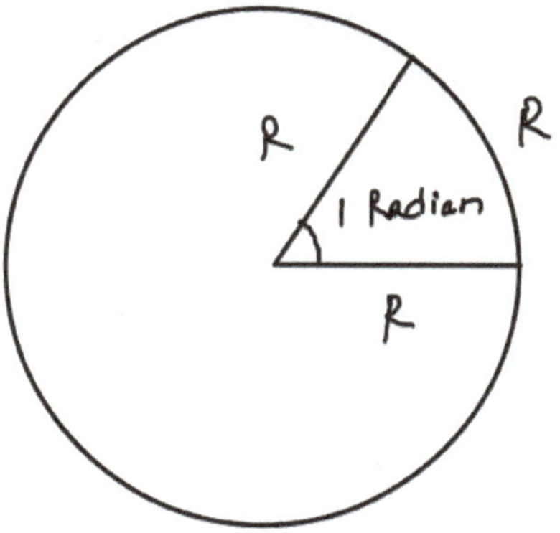

Figure 6.3.3

If one radian corresponds to the angle subtended by an arc of length *R*, what is the angle subtended by the entire circle, i.e., the angle subtended by an arc of length $2\pi R$? Let us try and derive that.

$$Angle\ subtended\ by\ arc\ of\ length\ R = \frac{Length\ of\ the\ arc}{Length\ of\ radius} = \frac{R}{R} = 1\ Radian$$

$$Angle\ subtended\ by\ entire\ circle = \frac{length\ of\ circumference\ of\ circle}{length\ of\ radius} = \frac{2\pi R}{R} = 2\pi\ Radians$$

From the above calculation, it might be clear to you that the $2\pi\ Radians$ equate to 360° or *360 degrees*. In other words:

$$1\ Radian \approx 57.3^\circ$$

Now that we have established the idea of a radian in two-dimensional space let us try and extend this idea to three-dimensional space. An angle (in degrees or radians) measures how two lines meet at a common endpoint in a two-dimensional space. Its equivalent in the three-dimensional space is called the *solid angle*. Informally, a solid angle is a measure of the amount of the field of view from some particular point that a given object covers. In other words, it is a measure of how large the object appears to an observer looking from that point. In figure 6.3.4, we have shown the field view from the center towards the surface of a sphere.

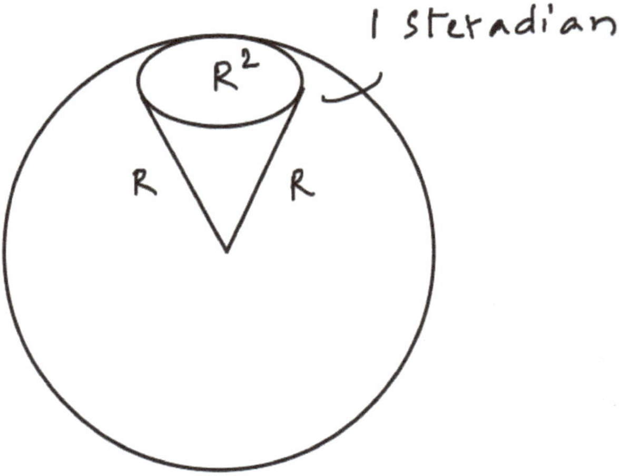

Figure 6.3.4

Similar to how we measure angles in radians, we measure solid angles in steradians. One steradian is defined as the solid angle of a sphere subtended (at the center) by a portion of the surface whose area is equal to the square of the sphere's radius. We can see this in figure 6.3.4.

$$Solid\ Angle\ subtended\ by\ surface\ of\ sphere\ of\ area\ R^2 = \frac{Surface\ area}{R^2} = \frac{R^2}{R^2} = 1\ Steradian$$

$$Solid\ angle\ subtended\ by\ entire\ sphere = \frac{Surface\ area\ of\ the\ sphere}{R^2} = \frac{4\pi R^2}{R^2} = 4\pi\ Steradian$$

The solid angle subtended by the entire sphere at the center is also the total solid angle about a point, which is *4π* steradians.

Why is the sphere's surface area (with radius *R*) $4\pi R^2$? Here is an interesting derivation. By integrating the circumference of a circle over its diameter, we will arrive at the surface area of a sphere.

Let us equate the diameter of the sphere of radius *R* to *D*. In other words, let *D=2R*.

Surface area of a sphere of radius R

$$
\begin{aligned}
&= Twice\, the\, surface\, are\, of\, a\, hemisphere\, with\, radius\, R \\
&= 2\int_0^D \pi x \\
&= 2\pi \int_0^D x \\
&= \pi D^2 \\
&= 4\pi R^2
\end{aligned}
$$

6.4 Symmetries in vector calculus

In this chapter, we shall explore some fundamental symmetries that exist in vector calculus. Such symmetries are called vector identities. In mathematics, an identity is an equality that relates one expression to another. We shall explore two fundamental symmetries that will come in handy in our journey to understand $e=mc^2$. The first symmetry is the one that exists among second partial derivatives, and we shall understand that through an example.

$$Let\,us\,consider\,a\,function\; f(x,y) = x^2 + 2xy + 3y^2$$

With two variables *x and y*, the second derivative of *f(x,y)* could be derived in multiple ways, as shown below:

$$\frac{\partial^2}{\partial x^2}, \frac{\partial^2}{\partial y^2}, \frac{\partial^2}{\partial x \partial y}, \frac{\partial^2}{\partial y \partial x}$$

We shall focus on the last two:

$$\frac{\partial}{\partial x \partial y} f(x,y) = \frac{\partial}{\partial x}(2x + 6y) = 2$$
$$\frac{\partial}{\partial y \partial x} f(x,y) = \frac{\partial}{\partial y}(2x + 2y) = 2$$

It is no coincidence that the above two values are the same. Not just for the above function, the second derivative of mixed partial equations is equal as long as the second derivative is a continuous function. This symmetry is the essence of Schwarz's theorem. We will not prove that here, but the catch with this symmetry is that the second derivative needs to be a continuous function. In our case, the second derivative is a constant, which is undoubtedly not a non-continuous function. What are non-continuous functions? In simple words, we cannot draw a non-continuous function on a graph without taking the pencil from the paper. We have plotted an example of a non-continuous function in figure 6.4.1. As you might observe, a small change around $x=1$ gives a huge difference in the value of y. The above symmetry does not apply to functions like these, and in our journey towards understanding $e=mc^2$, we shall consider only continuous functions.

$$y = \frac{1}{1 - x}$$

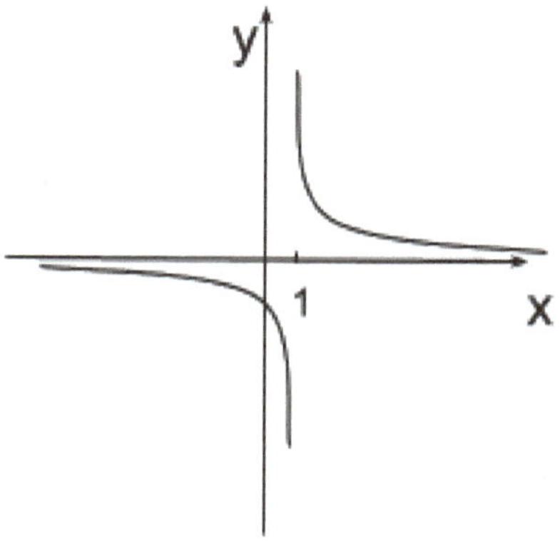

Figure 6.4.1

The second identity we shall explore is the *curl of a curl.* We have stated the identity below, and we shall be deriving the same in the rest of the chapter.

$$\nabla \times \left(\nabla \times \vec{E}\right) = \nabla\left(\nabla \cdot \vec{E}\right) - \nabla^2 \vec{E}$$

$$For\,a\,vector\;\vec{E} = u_1\vec{i} + u_2\vec{j} + u_3\vec{k} = \begin{bmatrix} u_1 \\ u_2 \\ u_3 \end{bmatrix}\;and\;\nabla = \begin{bmatrix} \frac{\partial}{\partial x} \\ \frac{\partial}{\partial y} \\ \frac{\partial}{\partial z} \end{bmatrix},$$

$$\nabla \times \left(\nabla \times \vec{E}\right)$$

$$= \nabla \times \left(\left(\frac{\partial}{\partial y}u_3 - \frac{\partial}{\partial z}u_2\right)\vec{i} + \left(\frac{\partial}{\partial z}u_1 - \frac{\partial}{\partial x}u_3\right)\vec{j} + \left(\frac{\partial}{\partial x}u_2 - \frac{\partial}{\partial y}u_1\right)\vec{k}\right)$$

$$= \left(\frac{\partial}{\partial y}\left(\frac{\partial}{\partial x}u_2 - \frac{\partial}{\partial y}u_1\right) - \frac{\partial}{\partial z}\left(\frac{\partial}{\partial z}u_1 - \frac{\partial}{\partial x}u_3\right)\right)\vec{i} +$$
$$\left(\frac{\partial}{\partial z}\left(\frac{\partial}{\partial y}u_3 - \frac{\partial}{\partial z}u_2\right) - \frac{\partial}{\partial x}\left(\frac{\partial}{\partial x}u_2 - \frac{\partial}{\partial y}u_1\right)\right)\vec{j} +$$
$$\left(\frac{\partial}{\partial x}\left(\frac{\partial}{\partial z}u_1 - \frac{\partial}{\partial x}u_3\right) - \frac{\partial}{\partial y}\left(\frac{\partial}{\partial y}u_3 - \frac{\partial}{\partial z}u_2\right)\right)\vec{k}$$

$$= \left(\frac{\partial^2}{\partial y \partial x}u_2 + \frac{\partial^2}{\partial z \partial x}u_3 - \frac{\partial^2}{\partial y^2}u_1 - \frac{\partial^2}{\partial z^2}u_1\right)\vec{i} +$$
$$\left(\frac{\partial^2}{\partial z \partial y}u_3 + \frac{\partial^2}{\partial x \partial y}u_1 - \frac{\partial^2}{\partial x^2}u_2 - \frac{\partial^2}{\partial z^2}u_2\right)\vec{j} +$$
$$\left(\frac{\partial^2}{\partial x \partial z}u_1 + \frac{\partial^2}{\partial y \partial z}u_2 - \frac{\partial^2}{\partial x^2}u_3 - \frac{\partial^2}{\partial y^2}u_3\right)\vec{k}$$

$$= \left(\frac{\partial^2}{\partial x^2}u_1 + \frac{\partial^2}{\partial y \partial x}u_2 + \frac{\partial^2}{\partial z \partial x}u_3 - \frac{\partial^2}{\partial x^2}u_1 - \frac{\partial^2}{\partial y^2}u_1 - \frac{\partial^2}{\partial z^2}u_1\right)\vec{i} +$$
$$\left(\frac{\partial^2}{\partial y^2}u_2 + \frac{\partial^2}{\partial z \partial y}u_3 + \frac{\partial^2}{\partial x \partial y}u_1 - \frac{\partial^2}{\partial y^2}u_2 - \frac{\partial^2}{\partial x^2}u_2 - \frac{\partial^2}{\partial z^2}u_2\right)\vec{j} +$$
$$\left(\frac{\partial^2}{\partial z^2}u_3 + \frac{\partial^2}{\partial x \partial z}u_1 + \frac{\partial^2}{\partial y \partial z}u_2 - \frac{\partial^2}{\partial z^2}u_3 - \frac{\partial^2}{\partial x^2}u_3 - \frac{\partial^2}{\partial y^2}u_3\right)\vec{k}$$

Applying the symmetry on second partial derivatives that we learned at the beginning of the chapter, we could rewrite the above equation as follows:

$$
\begin{aligned}
&= \frac{\partial}{\partial x}\left(\frac{\partial}{\partial x}u_1 + \frac{\partial}{\partial y}u_2 + \frac{\partial}{\partial z}u_3\right)\vec{i} - \left(\frac{\partial^2}{\partial x^2}u_1 + \frac{\partial^2}{\partial y^2}u_1 + \frac{\partial^2}{\partial z^2}u_1\right)\vec{i} + \\
&\quad \frac{\partial}{\partial y}\left(\frac{\partial}{\partial x}u_1 + \frac{\partial}{\partial y}u_2 + \frac{\partial}{\partial z}u_3\right)\vec{j} - \left(\frac{\partial^2}{\partial x^2}u_2 + \frac{\partial^2}{\partial y^2}u_2 + \frac{\partial^2}{\partial z^2}u_2\right)\vec{j} + \\
&\quad \frac{\partial}{\partial z}\left(\frac{\partial}{\partial x}u_1 + \frac{\partial}{\partial y}u_2 + \frac{\partial}{\partial z}u_3\right)\vec{k} - \left(\frac{\partial^2}{\partial x^2}u_3 + \frac{\partial^2}{\partial y^2}u_3 + \frac{\partial^2}{\partial z^2}u_3\right)\vec{k}
\end{aligned}
$$

$$
= \begin{bmatrix} \frac{\partial}{\partial x} \\ \frac{\partial}{\partial y} \\ \frac{\partial}{\partial z} \end{bmatrix} \left(\begin{bmatrix} \frac{\partial}{\partial x} \\ \frac{\partial}{\partial y} \\ \frac{\partial}{\partial z} \end{bmatrix} \cdot \begin{bmatrix} u_1 \\ u_2 \\ u_3 \end{bmatrix} \right) - \left(\frac{\partial^2}{\partial x^2} + \frac{\partial^2}{\partial y^2} + \frac{\partial^2}{\partial z^2}\right) \begin{bmatrix} u_1 \\ u_2 \\ u_3 \end{bmatrix}
$$

$$
= \nabla\left(\nabla \cdot \vec{E}\right) - \nabla^2 \vec{E}
$$

∇^2 is the Laplacian operator that expands to the following:

$$
\nabla^2 = \nabla \cdot \nabla = \begin{bmatrix} \frac{\partial}{\partial x} \\ \frac{\partial}{\partial y} \\ \frac{\partial}{\partial z} \end{bmatrix} \cdot \begin{bmatrix} \frac{\partial}{\partial x} \\ \frac{\partial}{\partial y} \\ \frac{\partial}{\partial z} \end{bmatrix} = \frac{\partial^2}{\partial x^2} + \frac{\partial^2}{\partial y^2} + \frac{\partial^2}{\partial z^2}
$$

6.5 Wave equation

We have spent a reasonable amount of time learning PDEs, i.e., partial differential equations, and in the wave equation, we will see its total utility value. The one-dimensional wave equation that we are about to derive has played a massive role in revealing the science behind light.

Here is the wave equation:

$$\frac{\partial^2 u}{\partial x^2} = c^2 \frac{\partial^2 u}{\partial t^2} \qquad (6.5.1)$$

If the above equation appears complex, you are not alone. Understanding the intuition behind this equation is not hard, and in this chapter, we shall unravel that mystery. Before we derive the above equation from scratch, let us try and understand how to create a wave using a string. You could try this with a rope or a string at home.

1. Connect one end of a string to a hook on the wall.
2. Hold it tight from the other end and oscillate, i.e., swiftly spin the end in either clockwise or counterclockwise direction. If you get the rhythm right, you will see the following pattern emerge.

We just created a standing wave, and by changing the speed of oscillation, we would be able to increase or decrease the number of nodes. A node is a point in the string where its vertical displacement is close to zero. An antinode is a point where the vertical displacement is the highest. The example shown in figure 6.5.1 has four nodes (N_1 through N_4) and three antinodes (AN_1 through AN_3).

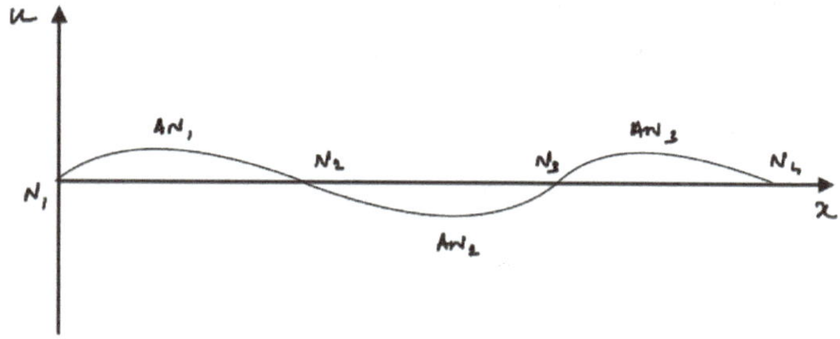

Figure 6.5.1

For the wave we just created, except for the nodes, you might observe the vertical displacement at every point in the string to vary over time. In other words, the vertical displacement can be expressed as the following function *u(x,t)*.

u(x,t) is the vertical deflection experienced at time *t* by point *x* of the rope

Before we derive the wave equation, let us try to understand the intuition of the wave equation. In our rope example, we created a wave by applying a specific force at the rope's end. As shown in figure 6.5.2, let us refer to the end of the rope as A. For any random point B on the rope to move, i.e., experience a vertical displacement, the force applied at A will have to pass through each infinitesimal part of the rope that lies between A and B. In other words, each infinitesimal part of the rope acts as a carrier of the force (created at A) which would also undergo a vertical displacement. In essence, the vertical displacement of point B at time *t* is dependent on the vertical displacement of every point in the rope that lies between A and B until time *t*. Mathematically, we can represent this relation as the following equation:

$$\frac{\partial^2 u}{\partial t^2} \propto \frac{\partial^2 u}{\partial x^2}$$

$$(or)$$

$$\frac{\partial^2 u}{\partial t^2} = C\frac{\partial^2 u}{\partial x^2} \ (6.5.2)$$

In the above equation, the component on the left-hand side is the rate of change of the vertical displacement at point B over time. It is nothing but a PDE of *u* keeping *x* constant. Similarly, the right-hand side is the rate of change of vertical displacement between A and B at a particular point in time. It is a PDE of *u* keeping *t* constant. Deriving the wave equation involves solving for constant *C* in equation 6.5.2.

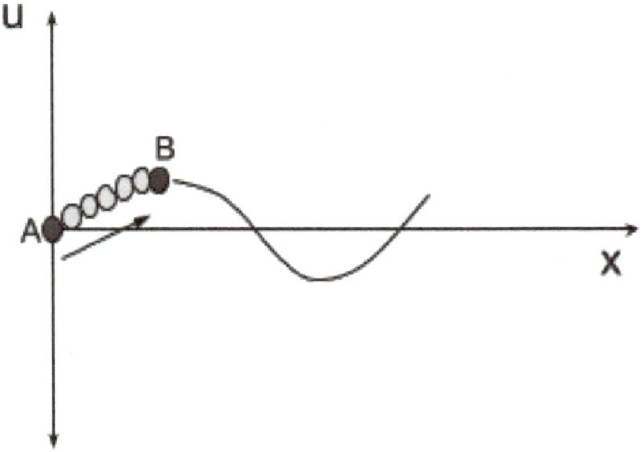

Figure 6.5.2

We are ready to begin deriving the wave equation, but we will have to make some assumptions about the string and wave we would create.

1. The string is of length *d,* and the mass of the string per unit length is constant. In other words, the density (per unit length) of the string ρ is a constant.
2. The string is perfectly elastic and does not resist bending.
3. The effect of gravitational force on the string is negligent.
4. The wave would cause the string to move up and down primarily. Its transverse movement, i.e., along its *x-axis,* is negligible.
5. The slope of the string is small at all locations. In other words, the vertical deflections are small.

Now, let us focus on an infinitesimal portion of the wave we considered in figure 6.5.1. We define that infinitesimal portion between points P and Q, as shown in figure 6.5.3.

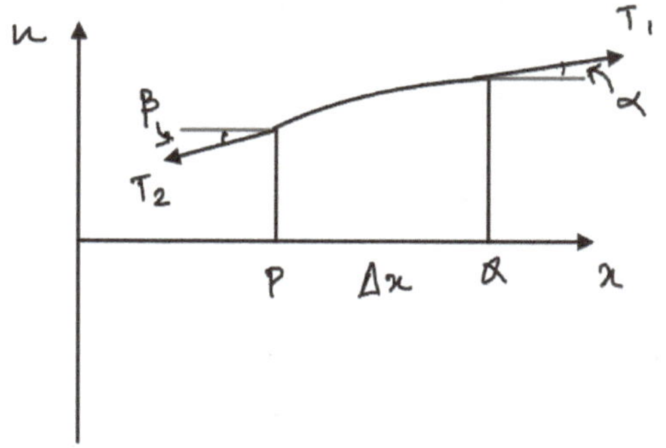

Figure 6.5.3

The string stays tight because, at every point, the string experiences a tension, i.e., an influence of two forces in opposite directions (left and right). Let us call these forces T_1 and T_2 as shown in figure 6.5.3. Since the string is curved while forming the wave, let α and β be the angles formed by the two forces with respect to the *x-axis*. For the fourth assumption to hold good,

$$T_1 \cos\alpha = T_2 \cos\beta$$

For the fifth assumption to hold good, the two angles α and β would be pretty close to each other and, more importantly, close to zero. We know that $\cos 0$ is 1. So the earlier equation could be approximated to T, as shown below:

$$T_1 \cos\alpha = T_2 \cos\beta \approx T$$

To form a wave, every part of the string would have to make an up and down movement. In other words, every part of the string would need to experience a vertical force, and by Newton's second law, we could equate it to the following:

$$\sum Vertical\,Force = ma_{vertical}$$

$$T_2 \sin\beta - T_1 \sin\alpha = m\frac{\partial^2 u}{\partial t^2}$$

At the nodes, the sum of the vertical forces would add to zero, and at the remaining points, the net vertical force will be non-zero. The mass of the string at an infinitesimal part is

$\rho\Delta x$. By replacing *m* with $\rho\Delta x$, we could rewrite the above equation as:

$$T_2 \sin\beta - T_1 \sin\alpha = \rho\Delta x \frac{\partial^2 u}{\partial t^2}$$

Now, by dividing both sides by *T*, we get:

$$\frac{T_2 \sin\beta - T_1 \sin\alpha}{T} = \frac{\rho\Delta x}{T}\frac{\partial^2 u}{\partial t^2}$$
$$\frac{T_2 \sin\beta}{T_2 \cos\beta} - \frac{T_2 \sin\alpha}{T_2 \cos\alpha} = \frac{\rho\Delta x}{T}\frac{\partial^2 u}{\partial t^2}$$
$$\tan\beta - \tan\alpha = \frac{\rho\Delta x}{T}\frac{\partial^2 u}{\partial t^2}$$
$$(\tan\beta - \tan\alpha) = \frac{\rho\Delta x}{T}\frac{\partial^2 u}{\partial t^2}$$

In the above equation, $\tan\alpha \, and \, \tan\beta$ are nothing but the slope at points P and Q.

$$\frac{(\tan\beta - \tan\alpha)}{\Delta x} = \frac{\rho}{T}\frac{\partial^2 u}{\partial t^2}$$

What we see on the left-hand side is the rate of change of slope, which we can rewrite as:

$$\frac{\partial^2 u}{\partial x^2} = \frac{\rho}{T}\frac{\partial^2 u}{\partial t^2}$$
$$\frac{\partial^2 u}{\partial t^2} = \frac{T}{\rho}\frac{\partial^2 u}{\partial x^2}$$
$$\frac{\partial^2 u}{\partial t^2} = C\frac{\partial^2 u}{\partial x^2}, \, where \, C = \frac{T}{\rho} \qquad (6.5.3)$$

We are almost there with our derivation of the wave equation. The only thing that is left to explain is, what does C denote?

Let us explore that through the units of the two components. We had considered ρ being the density per unit length of the string. The string we considered has constant mass throughout, and density is expressed in kg/m. On similar lines, T is the tension or force and is expressed in Newtons. We know one Newton equates to 1 m kg/s^2. With these, we can derive the unit for C to be m^2/s^2 or (m/s)^2. We know that m/s is the unit for velocity or speed. So, we can rewrite equation 6.5.3 as:

$$\frac{\partial^2 u}{\partial t^2} = c^2 \frac{\partial^2 u}{\partial x^2}, \; where \, c = \sqrt{C} \tag{6.5.4}$$

The above equation is the wave equation, and c corresponds to the speed of the wave.

6.6 Binomial theorem

Binomials are algebraic expressions involving two terms that are either added or subtracted. Here are some functions that are binomials.

$$f(a,b)=a+b$$
$$f(x)=x+2$$
$$f(x,y)=x^2+y^2$$
$$f(x,y)=x^2-y^2$$

As you might have noticed through the above examples, binomials are expressions involving two terms and not necessarily two variables. The following expressions are not binomials.

$$f(a,b,c)=a+b-c$$
$$f(x)=x$$

Let us try and expand a generic binomial expression (a+b) raised to different powers as shown below:

$$\begin{aligned}
(a+b)^0 &= 1 \\
(a+b)^1 &= a+b \\
(a+b)^2 &= a^2+2ab+b^2 \\
(a+b)^3 &= (a+b)(a^2+2ab+b^2) \\
&= a^3+3a^2b+3ab^2+b^3 \\
(a+b)^4 &= (a+b)(a^3+2a^2b+2ab^2+b^3) \\
&= a^4+4a^3b+6a^2b^2+4ab^3+b^4 \\
(a+b)^5 &= \ldots
\end{aligned}$$

You might observe a clear pattern to emerge for binomial expression with different exponents. The expansion of a binomial expression can be generalized as follows:

$$(a+b)^n = \sum_{k=0}^{k=n} \binom{n}{k} a^{n-k} b^k$$
$$where \binom{n}{k} = \frac{n!}{(n-k)!k!}$$

What we have stated above is the Binomial theorem. Let us see if we can generalize this a bit more. Let us start with expanding $(a-b)^n$

$$(a-b)^n = \sum_{k=0}^{k=n} (-1)^k \binom{n}{k} a^{n-k} b^k$$
$$where \binom{n}{k} = \frac{n!}{(n-k)!k!}$$

Applying the binomial theorem for $(a-b)^3$, we get the following:

$$(a-b)^3 = (-1)^0\left(\frac{3!}{3!0!}\right)a^3b^0 + (-1)^1\left(\frac{3!}{2!1!}\right)a^2b^1 + (-1)^2\left(\frac{3!}{1!2!}\right)a^1b^2 + (-1)^3\left(\frac{3!}{0!3!}\right)a^0b^3$$
$$(a-b)^3 = a^3 - 3a^2b + 3ab^2 - b^3$$

Before we go and perform our more significant generalization, let us first expand our binomial expansion for $(a-b)^n$.

$$
\begin{aligned}
(a-b)^n &= \sum_{k=0}^{k=n} (-1)^k \binom{n}{k} a^{n-k} b^k \\
&= (-1)^0 \left(\frac{n \cdot (n-1)(n-2)..1}{n \cdot (n-1)(n-2)..1 \cdot (1)} \right) a^n b^0 + (-1)^1 \left(\frac{n \cdot (n-1)(n-2)..1}{(n-1)(n-2)..2 \cdot 1 \cdot (1)} \right) a^{n-1} b^1 + \\
&\quad (-1)^2 \left(\frac{n \cdot (n-1)(n-2)..1}{(n-2)..3 \cdot 2 \cdot 1 \cdot (2 \cdot 1)} \right) a^{n-2} b^2 + \ldots + (-1)^n \left(\frac{n \cdot (n-1)(n-2)..1}{1 \cdot (n \cdot (n-1) \cdot (n-2)..1)} \right) a^0 b^n \\
&= (-1)^0 a^n b^0 + (-1)^1 (n) a^{n-1} b^1 + (-1)^2 \left(\frac{n \cdot (n-1)}{2!} \right) a^{n-2} b^2 + (-1)^3 \left(\frac{n \cdot (n-1)(n-3)}{3!} \right) a^{n-3} b^3 + \ldots + (-1)^n a^0 b^n \\
&= a^n - (n) a^{n-1} b^1 + \left(\frac{n \cdot (n-1)}{2} \right) a^{n-2} b^2 - \left(\frac{n \cdot (n-1) \cdot (n-2)}{6} \right) a^{n-3} b^3 .. + (-1)^n b^n
\end{aligned}
$$

Could we use the binomial expansion to solve for expressions that have a negative exponent? How would we expand the following expression?

$$\frac{1}{(a-b)^n} = (a-b)^{-n}$$

We can apply the idea of binomial expansion with a caveat. We will not be able to get the exact value but an approximated value. The reason is, when *n* is less than zero, our values of *k* will end up expanding from 0 to infinity (instead of *n*). We could write our generalized binomial expansion for negative and fractional exponents as follows:

$$(a-b)^{-n} = (-1)^0 a^{-n} b^0 + (-1)^1 (-n) a^{-n-1} b^1 + (-1)^2 \left(\frac{-n \cdot (-n-1)}{2!} \right) a^{-n-2} b^2 + (-1)^3 \left(\frac{-n \cdot (-n-1)(-n-3)}{3!} \right) a^{-n-3} b^3 + \ldots..$$

7. INDEX

ABOUT THE AUTHOR

Ramaswami Mohandoss was born and raised in Chennai, India. Programming was the first love that pulled him into Computer Science during his teens. He is now a seasoned technologist with over two decades of Software Engineering experience. He has authored over twelve white papers in Data & AI and has filed two patents. His love for programming got him interested in mathematics, and his interest in math led him to the world of physics. In this book, he explains and derives Albert Einstein's most famous equation from scratch.